GLOSTER

JAVELIN

The first multi-jet Delta all weather fighter, the Gloster G.A.5, one of the new types produced for the Royal Air Force fitted with

ROTOL

ACCESSORY DRIVE
EQUIPMENT

ROTOL LIMITED · GLOUCESTER · ENGLAND

GLOSTER

JAVELIN

THE RAF's
FIRST DELTA WING
FIGHTER

RICHARD A FRANKS

DALRYMPLE
& VERDUN
PUBLISHING

The Gloster Javelin
The RAF's First Delta Wing Fighter
Richard A Franks

ISBN 1-905414-02-1

First published in 2006 by
Dalrymple & Verdun Publishing
33 Adelaide Street, Stamford,
Lincolnshire PE9 2EN
Tel: 0845 838 1940
mail@dvpublishing.co.uk
www.dvpublishing.co.uk

© Concept and design
Dalrymple & Verdun Publishing and
Stephen Thompson Associates

© Richard A Franks 2006
© Richard J Caruana - Colour Profiles

Printed in England by
Ian Allan Printing Ltd
Riverdene Business Park, Molesey Road
Hersham, Surrey, KT12 4RG

Acknowledgments
A big word of thanks must go to Barry
Jones for all his assistance, support and
insight, without which this book would not
have happened. His knowledge of the
subject assisted me in so many ways and
I am indebted to him for the selfless
manner in which he answered my
questions and helped me locate
photographs etc. Thanks also to Derek
James, who was able to supply me with
many of the images you see in this title;
Richard J Caruana for all his support and
colour profiles, Godfrey Mangion, Ray
Deacon and Martyn Chorlton for helping
me with photographs. Further thanks
to Martin Derry for his invaluable
contribution.

**The publishers are greatly indebted
to Roger Lindsay and gratefully
acknowledge his contribution in
helping to edit this title.**

Title page: *Four Javelins F(AW) Mk Is
(XA626, XA627, XA623 and XA572)
from No.46 Squadron formate for
the camera. Note the three styles of
squadron badge presentation,
suggesting that this view was
possibly taken in 1956/57.*

Opposite page: *The old and the new
at RAF Coltishall in February 1961.
The Lightning era beckons though
the Javelin (XH894) has a little
longer to serve in the UK yet.*
Both Martyn Chorlton

CONTENTS

	Preface	
Chapter 1	Deciding on the Delta	8
Chapter 2	Testing, Spinning and Super-stalling	14
Chapter 3	The 'Flat-iron' goes into Service	22
Chapter 4	The Javelin Abroad	42
Chapter 5	Javelin Technical Description & Armament	48
Chapter 6	Projects, Proposals and Drawing Board Javelins	66
Chapter 7	Camouflage & Markings	76
Appendix I	Technical Data	98
Appendix II	Javelin Squadrons	100
Appendix III	Javelin Production	102
Appendix IV	Glossary and Bibliography	111
	Index	112

NOTE
Please note regarding designation and terminology. In
researching this book I have discovered a wealth of
different nomenclature for the Javelin marks. In many
cases a consistent use of the term F(AW) has been
used, while in others a mix of F(AW) and FAW are
present and in some instances just FAW has been used.
The main reason for this I think stems from the official
terminology by the Air Ministry and (later) Ministry of
Defence. To quote these sources the 'correct'
designations are as follows; F (AW) Mk 1, FAW Mk 2,
T Mk 3, F (AW) Mk 4, FAW Mk 5, FAW Mk 6,
FAW Mk 7, F (AW) Mk 8 and F (AW) Mk 9. For this title
however I have opted for consistency and will use F(AW)
throughout.

In addition to the inconsistencies mentioned above
variation also applies to some of the historical
information contained in this publication. Undoubtedly
this is caused by gaps in the public record or human
error after the passage of fifty years and more. The
publishers would welcome any new information or
correction that readers may be aware of by either
forwarding a letter to Dalrymple & Verdun, or by visiting
our forum at the following web address
www.dvpublishing.co.uk/forum

FOREWORD

There are many others far better qualified than myself to write the foreword to this lavishly illustrated history of the Javelin. I know of some, within the aircraft's comparatively short in-service life, who flew this unique aeroplane for at least three consecutive tours serving in the UK, the Middle East and finally the Far East during confrontation with Indonesia. Perhaps they will generously forgive the early intrusion to this work of a one-tour man whose only distinction in the history of the Javelin was to be the 'first-tourist'.

Reading Richards Frank's detailed history of the development of the Javelin, its introduction to service and its role in the generation of national air power in the 1950's and 1960's has brought back many happy memories. His book also serves as a timely reminder of the huge contribution British test pilots made to our security in the darkest days of the Cold War. Chapter 2 pays a detailed and proper tribute to their courage and skills throughout the rather troublesome development phase of the Javelin before its delivery to the first operational squadron – its compulsive reading to say the very least.

For my part I flew the Javelin F(AW) Mk 5 at the OCU at RAF Leeming and the F(AW) Mk 7 and F(AW) Mk 9 on No.64 Squadron, first at RAF Duxford with subsequent moves to RAF Waterbeach and RAF Binbrook. Having enjoyed the pleasure of flying many other aircraft types since those days, I can say without fear of contradiction that none exhibited the totality of eccentricities which came with the Javelin. Starting the engines could be exciting, n.b. the comment in Chapter 5 which is something of an understatement! I still vividly recall the scene at Waterbeach on Friday 13th April 1962 when Flt Lt Bob Lockhart's F(AW) Mk 9 exploded on start-up which cat 5'd the aircraft and caused severe collateral damage to adjacent aeroplanes – as well as leaving Bob and his navigator, Paddy Hamilton, with concussion. After 64 Squadron, I was never again to fly an aircraft which you had to spin in order to recover from a stall, nor one within which selection of reheat at lower altitude caused a reduction in thrust. On the other hand, selection of full airbrake at high speed certainly required a timely warning to the navigator; the deceleration was truly fierce and unless pre-warned the nav was likely to suffer some rearrangement of his nose if his eyes were glued to the radar screen. But oddities apart, the Javelin was a most effective night-fighting platform with crew training focussed from the start on the visual identification at night of targets. To get in close on a truly black night against an evading target required the very best teamwork of pilot and navigator. And although the advent of the Firestreak air-to-air missile markedly increased the killing potential of the Javelin, few if any Javelin aircrew of my time will forget the 'stimulation' of a snap-up guns attack designed to overcome the rearward firing radar-laid guns in Soviet bombers of that era.

In writing this foreword, many other memories come flooding back – the introduction of the in-flight refuelling and my first long-range overseas flight; quick reaction air-defence alert on two minutes cockpit standby with 1¾ hours strapped in and 1¾ hours out alternating through the long winter nights; and then the compensation of a three months summer detachment to Cyprus for air gunnery as well as air defence duties. But I recall above all the debt I owe to so many friends on No.64 Squadron whose instruction and encouragement got me safely through my first and quite unique flying tour.

And of those I acknowledge with particular gratitude Flt Lt David Holes. We crewed up together on the OCU. He alone was prepared to risk everything in the hands of an experimental first tourist and we remained together until I was posted to Hunters, my original destination after completing flying training. Dave's patience, his good humour and the skills he had accumulated on two previous night-fighter tours were a constant throughout. I never cease to wonder how Dave and his fellow nav rads (navigator/radar operators) so stoically accepted the quirks of their pilots as well as the peculiarities of a unique aeroplane.

Air Chief Marshal
Sir Richard Johns GCB CBE LVO
Windsor, March 2006

*Gloster 'Javelin' prototype.
From a postcard printed in
1952 by B Mathews
(Photo Printers) Ltd
Bradford. Card No 343.*

*WT841. The prototype
T Mk 3, first flown in
August 1956.
Both Martyn Chorlton*

Chapter 1: **DECIDING ON THE DELTA**

To fully understand the Javelin you need to put it within the context and political situation that existed at that time. The immediate post-war period was one that saw a massive leap in aeronautical development and one that was coupled with the underlying threat perceived from the Eastern Bloc. Memories of the dropping of the atomic bombs on Japan were still very vivid and the threat posed by long-range, high-flying bombers was being taken very seriously in most western nations. In the UK the need for effective fighter precautions against the envisaged nuclear bomber led to the issuing of two specifications, F.43/46 for an interceptor and F.44/46 for a two-seat fighter. Many previous published works on the Javelin cite these as the starting point for the type, while others quote previous development on high-speed versions of the existing Gloster fighter, the Meteor. The truth is that both are valid comments, as the type was designed to meet a specification, but of course it drew on all the previous design studies that Gloster had undertaken in relation to a high-speed fighter. The first of the two above specifications called for an aircraft capable of intercepting and destroying an enemy bomber flying at high altitude, while the latter specified that this task was to be achieved at 40,000ft with a speed of no less than 540kts so that a bomber flying at 480kts could be caught. A service ceiling of 45,000ft was envisaged and the aircraft had to be at altitude within 10 minutes of the pilot starting the engines. Initial armament was to have been the recoilless gun developed at Fort Halstead. This weapon used the simple recoilless system whereby as the 3.7in or 4.5in projectile was fired out of one end of the barrel, a similar weight was pushed out the back of the barrel thus removing the stresses on the airframe associated with the firing of big calibre weapons. These weapons could not be reloaded in the air, so a barrage of four or six per wing was initially called for. Fortunately guided missiles were being successfully developed at the same time, and when sufficient progress with these weapons was achieved the recoilless system was abandoned and on-board weapons in the form of four 30mm ADEN cannon (developed from the German WWII era Mauser MG 213C) were installed.

Throughout 1947 Gloster prepared various design proposals to meet the specifications. This resulted in a mass of P-series design projects and a full list of those associated with the Javelin are listed in Chapter 6. Much of the design work carried out by George Carter and his design team at Gloster, centred around developments that had a distinct 'Meteor' look to them, as well as many using a delta layout. Much research into deltas was available to the Allied powers

---

SPECIFICATION F.44/46

Type: Two-seat fighter.

Role: Daylight interception and destruction of high speed, high altitude bombers.

Interception Height: At least 40,000ft.

Maximum Speed: At least 540kts to intercept bomber travelling at 480kts.

Service Ceiling: 45,000ft.

Rate of Climb to 45,000ft: Less than 10 minutes from start-up.

Minimum Endurance: 2 hours (from take-off to landing, including taxying, climb to 25,000ft and 15 minutes combat at altitude).

Take-off Distance: 1,500 yards.
Landing Distance (clearing 50ft obstacle): 1,200 yards.

Manoeuvre Stress Levels: 4g at top speed at sea level.

Artificial Pressure System: Reproduce equivalent pressure for 25,000ft at 45,000ft actual height.

Radio & Radar: VHF radio, Airborne Interception (AI) radar, navigational aids and (possibly) blind landing equipment.

Oxygen System: Supply for 2.5 hours at 25,000ft.

Safety Equipment: Crew ejection seats, jettisonable canopy, dinghies and parachutes.

Airbrake Operation: 4 seconds.

Armament: Initially this called for the 3.7in and 4.5in recoilless guns, but this was revised to call for four 30mm cannon with sufficient ammunition for 15 seconds firing per gun.

Production: Economic production rate of 150 units at a maximum rate of 10 per month.

---

in the immediate post-war period as this had been captured from German manufacturers at the end of the war. Much of this research had been carried out by Dr Alexander Lippisch and some of the Gloster proposals drew on this information. Although the initial design proposals from Gloster had drawn on the Meteor lineage, with the likes of the P.228 and P.234, by August 1947 the design concepts had centred around the P.248 and P.250 designs. The P.234 proposal had already been identified as superior during

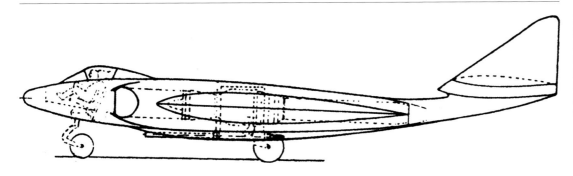

This line art produced by Gloster at the time shows Project Number 234, which was evolved into P.259 and thus on to P.272 which started to look much like the G.A.5 and therefore the Javelin.

the Tender Conference on the 9th April 1947, but although Gloster received instructions to go ahead with the project in June, no order was placed. Gloster therefore amended the P.248 and P.250 to incorporate those aspects that were approved of in the previous projects and then were informed that the Air Staff had changed their mind on the actual requirements. The Ministry of Supply revised the specification and asked that the new Red Hawk air-to-air missile system be installed along with a large Airborne Interception (AI) radar system. To this end Gloster amended the two previous projects to meet these new demands, thus offering the P.259. For the first time, the adoption of a two-seat configuration was necessary due to the new large AI radar system.

Elsewhere other manufacturers were also working to meet the changing specifications, with Hawker producing the P.1057 and De Havilland the DH.110. Hawker later went on to move away from their two-seat P.1057 and develop it as a single-seat fighter in the form of the P.1067 that became the Hunter. De Havilland on the other hand continued with their work on the DH.110. All of these various design proposals were of interest to the Air Ministry, but post-war development and evolution of RAF requirements were such that by 1948 the initial specifications were superseded by new ones that encompassed all of the new technical requirements that had been requested in the proceeding year. The new specifications were thus F.3/48 for the day fighter and F.4/48 for the night

The Boulton Paul P.120, VT951, was built to investigate delta-wing handling when the aircraft was fitted with a tailplane. The trials programme was cut short on 29th August 1952, when the aircraft crashed after only 11hr 15min total flying time.

Having made its maiden flight on 4th September 1949, the Avro 707, VX784, was shown in the static park at that year's SBAC Display.
Both Barry Jones

A 'family photo' taken at Farnborough's 1953 SBAC Display, as the 707B VX790 leads the procession, followed by 707A WZ736, 707C WZ744, and 707A WD280 bringing up the rear.

The one and only Avro 707C, WZ744, first flown on 1st July 1953, was unique in being the only British single-engined, two-seat delta wing aircraft. Both Barry Jones

fighter and the Air Ministry was hoping that a single design could be used to meet the requirements of both specifications. At Gloster two distinct new design proposals were prepared. The first, P.272 (developed from the P.259), had a delta wing and tail, while the latter P.275 was an unusual delta 'dart' design that saw the pilot situated in the leading edge of the long vertical fin fillet. This design will be familiar to many as it bore a striking resemblance to the Lippisch-designed WWII-era P.13a project which was intended to be powered by a ram jet fuelled with 'foam coal' but which was never actually built or flown. Both of these designs featured rotating wing tip controls instead of ailerons and this was simply because little information was available on lateral control in relation to the delta wing planform. Both designs were to use the Rolls-Royce Avon engine, the former using two and the lat-

ter just one. Thanks to research in the USA about aileron controls with high-speed swept wing designs like the North American F-86 Sabre, the rotating tip controls were soon dropped in favour of normal ailerons. In January 1948 George Carter had moved to became the Technical Director at Gloster, so his place in the design team was taken by Richard W Walker. The rest of the design team was Mr I James as Chief Technician, D I Husk as Chief Aerodynamicist, J F Cuss as Chief Stressman with P G Crabbe in overall control as Managing Director. At this time the decision was also made to concentrate all efforts on meeting specification F.4/48, with all work on F.3/48 being abandoned. Various new design proposals were now developed and these used one, two, three or four Avon engines and there was one with a single Avon engine and two rocket engines giving 3,000lb of

WHY A DELTA?

Considering the comments made elsewhere about the lack of knowledge in certain areas of the flight envelope for the Delta planform, why did Gloster seem so taken with it? The main benefits are as follows.

1 The high taper ratio offered a low aspect ratio.

2 The sharp sweep of the leading edge was good at high speed.

3 The large wing area helped the type achieve a short take-off run and performance at altitude.

4 The long root chord gave a thin wing which also offered large internal volume for fuel, armament and a retractable undercarriage.

5 The deep wing chord lead to a smooth blending with the fuselage; the fuselage providing 1/5th of the total lift.

6 All the above combined to give low structural weight and maximum integral strength for both bending and torsional stresses.

thrust; all of them being of delta planform. Much research work in the wind tunnel was undertaken and this resulted in many detail changes to the initial proposals. One of the main problems though remained, and this was the lack of knowledge and experience with the delta, especially through the stall and at low speeds. As a result of this the Air Ministry started a full research project into the delta wing planform. These tests, undertaken in co-operation with the RAE at Farnborough, led to the building of the Boulton Paul P.111. This type did little to help with the Gloster design though, as it did not fly until October 1950. The delta 707A and 707B built by Avro were not ready either in time to benefit Gloster, as the first did not fly until the 6th September 1950. In conjunction with the delta planform Gloster opted to include a high-set delta tailplane. This unit was mounted on top of the vertical fin to cope with the various trim changes that were anticipated in relation to the speed range of the new design. In the original design the tailplane was a slab control with no elevators, but the research with the F-86 in the USA confirmed that elevators were

effective for transonic control. As a result the tailplane was modified to be trimable with conventional elevators. The separate tail also gave the added benefit of allowing the incorporation of flaps in the inboard sections of the wings and thereby removing the need for the high angle of attack associated with landing a delta. The initial choice of powerplants for the new design was between two 6,500lb thrust Rolls-Royce AJ.65 Avons or two 7,000lb thrust Metrovik (later Armstrong-Siddeley) F.9s (later to become the Sapphire). By July 1948 Gloster were proposing the use of two of the Sapphire 2 engines that were projected as giving 9,000lb of thrust, but this was revised back to the standard Sapphires of 7,500lb thrust each for project P.280. In the end the Sapphire was chosen instead of the Avon simply because of the thrust benefit and in August 1948 Gloster submitted their final design brochure.

DH.110, the alternative 'safe guard'

As has already been mentioned, at the same time as Gloster was working on the design that would lead to the Javelin, De Havilland were also working on a similar design study. The Royal Navy issued specification N.40/46 in January 1947 for a two-seat night fighter and it is not surprising to see that the Air Ministry specification F.44/46, which was issued on the 24th January 1947, was very similar in many respects. De Havilland opted to submit a single proposal, under company designation DH.110, for both specifications. The type featured a fuselage 'pod' like the Vampire with twin booms, though with two jet engines, a two man crew and swept wing planform. Like Gloster, De Havilland had the benefit of ex-German aerodynamic research and it opted to build the DH.108 Swallow to research the effect of swept wings on high-speed designs. From the research with the DH.108, De Havilland were able to select a sweep angle of 40° and to give added safety in a single engine failure both engines were fitted side-by-side in the fuselage pod. As with the Gloster design, the De Havilland proposal had four 30mm ADEN cannon fitted, although in the DH.110 these were installed below the cockpit in the fuselage, thus allowing the wings to be free of guns and their associated ammunition and feeds. Other than the naval equipment such as wing fold, installed to meet Naval Specification N.40/46, the DH.110 was

The first DH.108, TG283, made its maiden flight, lasting 30 minutes, from Woodbridge in Suffolk, on 15th May 1946 and had its leading edge Handley Page slats locked in the open position, as its role was for low-speed handling trials. Barry Jones

The first DH.110, WG236, made its maiden flight on the 26th September 1951. Barry Jones

designed to meet the stringent requirements of F.4/48 which in turn had been created to meet the Operational Requirement (OR) 227 of the Air Staff issued in December 1946.

The DH.110 was of interest to the RAF and in April 1949 the Ministry of Supply (MoS) placed an order with De Havilland for nine prototypes that were to be built as seven night fighters and two long-range fighters. The production of the type was only ever seen as a safety measure should the Gloster design (designated the GA.5) be a failure. In the end only five DH.110s were built for the RAF requirement (F.4/48) and the other four were diverted to the Royal Navy in relation to Specification N.14/49. In November the Admiralty decided that the DH.110 was too costly and risky and that they would not wait for it, so they opted for the less risky DH.112 Sea Venom instead and the prototype DH.110s for N.14/49 were cancelled. (See sister volume on the DH Sea Vixen.) Interest in the DH.110 from the RAF in regard to F.4/48 was also to decline, and at the same time as they reduced their commitment to the Gloster GA.5 prototypes they did the same with the DH.110 and so just two (WG236 and WG240) of the night-fighter version were required and the order was reduced accordingly. On the 7th July 1952 the Ministry of Supply announced that the Gloster design had been adopted and all RAF interest in the DH.110 came to an end.

The Gloster GA.5

On the 13th April 1949 both De Havilland and Gloster received instructions from the Ministry of Supply to proceed with the construction of four prototypes for flight testing and one airframe for structural testing. Four prototypes were considered to be too few to

develop such advanced designs, but this situation worsened in November 1949 when the Labour Government reduced the order to just two of each! This was rectified eventually, but it took until March 1951 before the order was increased again to five airframes plus one for development of a trainer version. In the intervening seventeen months much momentum on the design had been lost and this was to be a deciding factor that led to the delay in the arrival of the type in RAF service. The delay was also to see the first production airframe being manufactured before the prototype airframes were completed.

Once the 'Order to Proceed' was received from the Ministry of Supply in April 1949 construction of the prototype began at Gloster's experimental department at Bentham. By February 1950 the main structural design work had been done and by June full systems drawings had been completed as well. By June 1951 construction of the prototype was complete. The airframe was very similar to the P.280 proposal and was fitted with a solid metal nose cone in lieu of the radar that would be installed in production airframes. Initially the prototypes had the twin cockpit partially skinned, so that the radar operator in the rear seat was in almost complete darkness. This was similar to the set-up seen in the DH.110, with the radar operator situated in the fuselage alongside and below the pilot. Neither type allowed any outside light to enter this crew station, as it was initially felt that the radar operator would work more effectively in a low-light environment, although the GA.5 was later to have a reprieve and see the entire cockpit glazed.

The first prototype, WD804, was moved by road from Bentham to the Flight Test aerodrome at Moreton Valence in July 1951 and a two month period lapsed while the type was reassembled and prepared for its first flight. Apart from the usual test equipment an anti-spin parachute unit was fitted on top of the tailplane. This airframe was just a flying shell, for evaluation purposes only and it was not fitted with any military equipment (the lack of radar weight being compensated by the solid metal nose cone). The initial programme of tests were all ground based and during one test run one of the engine intakes partially collapsed. On further static tests it was found that the low pressure within the intake during engine running was higher than envisaged and this had caused the intake to partially fail. To overcome this all the stiffening rings on the outside of the intake duct were dou-

bled and no such further problems were encountered. Flight testing was undertaken by Gloster's Chief Test Pilot, Sqn Ldr W 'Bill' A Waterton AFC, although in the month after reassembly the aircraft was just to undertake ground running and taxying trials. These tests saw about 30 high speed runs down the runway at Moreton Valence and as a result of all these preliminary tests a number of modifications were incorporated. With all of these tests completed, the modifications embodied and the weather right, on the 26th November 1951 WD804 with Sqn Ldr Waterton at the controls took to the air for its first flight.

DH.110, WG240 seen at Farnborough in 1953, while Javelin prototype WT827 is parked alongside.

The first published photograph of the Gloster GA.5 prototype, taken before it had received national markings and its WD804 serial number.
Both Barry Jones

Chapter 2: **TESTING, SPINNING AND SUPER-STALLING**

Bottom: *Prior to its maiden flight from Moreton Valence on 26th November 1951, flown by Sqn Ldr 'Bill' Waterton, WD804 shows the original jet pipe fairing and the metal rear section of the cockpit canopy, with its two small round portholes, covering the radar operator.*

Opposite page top: *Severe rudder vibration encountered during the 34 minute maiden flight, was traced to interference in airflow at the rear fuselage reacting with the jet efflux. One of the early changes made to the rear-end profile is evident here.*

Opposite page bottom: *In a further reshaping of WD804's rear-end, the oil-vent pipes running from the Sapphires' centre and rear bearings were extended.*
All Derek N James

The first flight of the prototype lasted for 34 minutes and did not go completely smoothly. As soon as Bill Waterton had got the aircraft into the air, vibrations started throughout the airframe, as the speed increased so did the vibrations. By careful deduction Bill Waterton realised that the problem related to the rudder and the pilot of the accompanying Meteor chase plane was able to confirm this, as he could see the rudder being buffeted. As a result of this phenomenon Bill Waterton decided to undertake a relatively limited assessment of the aircraft in this initial flight and reduced the speed accordingly before returning to Moreton Valence. After the flight the airframe was inspected and the path of oil on the lower rear fuselage (which came from the 'total loss' lubricating system employed on the Sapphires centre and rear bearing assemblies) clearly showed the path of the disrupted airflow. Investigation into the cause of this was undertaken by Gloster in both the wind tunnel and through further flight testing. For the flight tests wool tufts were attached to the rear fuselage of the airframe and their behaviour in flight was noted. The results of this showed that there was vertical airflow over certain parts of the fin and this was determined to be caused by the merging of the jet efflux and the air around the rear fuselage, which disturbed the airflow across the fin causing disturbance around the rudder and in turn resulting in it being buffeted. The initial fix for this problem was to lengthen the fairings around the jet exhausts to bring them level with the extreme tip of the fuselage. Three such extensions were carried out in gradual succession, each giving an extra 50kts before

the buffeting set in, but the problem was not completely overcome. Further tests ensued and it was determined that the guard rails fitted around the anti-spin parachute mounted on the tailplane were the cause and once these were removed the buffeting did reduce sufficiently for the test programme to resume. The problem was finally eradicated entirely in March 1952 with the extension of the rear fuselage in what is termed as the 'pen nib' fairing and at the same time the jet exhausts were also lengthened by 2ft.

With the problem overcome by the initial 'fix', the flight testing programme for the prototype resumed. As much of the flight envelope that could be investigated was, with speeds being gradually increased. Once the airflow problem was finally fixed in March 1952, Gloster issued their initial flight assessment of the type to the Ministry of Supply. The assessment was very positive, as the type had handled well and the systems, airframe and engines had been exceptional in their serviceability. The flight range had been conducted at altitudes from sea level to 43,000ft and a top speed of Mach 0.915 had been achieved in level flight at just 60% of maximum engine power. With the adoption of the revised new fuselage and extended jet exhausts, the fitment of more powerful Sapphire engines and with the anti-spin parachute removed the prototype started the second phase of the test programme in April 1952. It was at this time that the Ministry of Supply instigated a series of comparison trials with the GA.5 and De Havilland DH.110. During these trials it became obvious that the Ministry saw greater potential in the delta configuration of the GA.5 than in

WT827 during an early test flight.

that of the twin-boom DH.110, even though the latter design had the better performance of the two and was easier from a production viewpoint. It was felt by both the Ministry and the RAF assessment teams that the potential for the delta design was much higher and thoughts were also expressed that a thin wing version would be a real option at a future stage.

The test programme continued without a hitch until the 99th flight on the 29th June 1952. This flight was to investigate the handling of the aircraft at high speed and low altitude because of the continual revision of the flight controls that were on-going at this time. Bill Waterton took the aircraft up and established it for a high speed flight above the Cotswolds at 3,000ft. The first run went without a hitch, so he turned over Oxford and headed back. As he was approaching Brize Nor-

ton the aircraft exhibited, without warning, severe vibration. The vibrations lasted no more than two and a half seconds and were followed by two loud bangs. These bangs were caused by both elevators breaking away, as the vibrations had been caused by severe elevator flutter. After the initial shock and confusion Bill Waterton was able to obtain some form of control and slowly gained height to 10,000ft while also slowly reducing the speed to 300kts. At this stage he was considering that he would have to abandon the aircraft and let it crash in the Bristol Channel, so a course was set for this area. Using his skills he was able to gain some level of control, even without elevators, by using the variable-pitch tailplane. This was electrically operated and although the operation of the servos that moved it were slow, some control was achieved. This

The third prototype, WT827, in flight. You can clearly see the covered-in rear cockpit area and the clear cover over the DF loop in the back of the canopy 'hump'.
Both via author

By the time the third prototype, WT827, had made its maiden flight on 7th March 1953, the name Javelin had been bestowed on the GA.5. WT827 was the first prototype to have the wing-mounted four-gun armament and British AI.17 radar installed within a revised nose cone profile. The radar operator had also received an additional window in the top of his canopy.

During the programme to determine the optimum nose-cone shape on the third prototype, this bull-nosed version was test flown during 1953. Both Barry Jones

was not easy as the control for the tailplane was in the form of a conventional trim wheel that was linked to a repeater system; something that Bill Waterton himself had insisted be installed. Happy that he had sufficient control using this system, he decided to try and land the aircraft and therefore headed for A&AEE Boscombe Down with its very long runway. A number of dummy approaches were made to the airfield until he was happy that he could control the aircraft in the descent and deceleration, although he had opted to do this at two-thirds above the normal recommended landing speed. He made his final touch-down on the long runway, but the higher landing speed coupled with the high rate of descent due to the lack of full (conventional) control resulted in a very hard landing and the aircraft took to the air once again before bouncing several times more (some eyewitnesses state that one of these 'bounces' was as high as the nearby hangar!). This heavy bouncing beat the aircraft to destruction and on the last but one bounce, the port undercarriage was pushed through the wing, puncturing the fuel cell as it went. The surface abrasion caused spilling fuel to ignite and the aircraft slewed off the runway, coming down for the final time and push-

ing the starboard undercarriage through the wing and puncturing that fuel cell. The aircraft was a mass of flames, and Bill Waterton found to his horror that he could not get out, as the canopy was jammed and only a desperate effort saw him smash it and force it back so that he could get out. With the arrival of the fire engines, he instructed them to direct their foam onto the nose and tail sections, to try and protect the flight recording equipment that was located there. Once the fire was extinguished he was rewarded by being able to recover some of the auto-observer records. Later he was to be awarded the George Medal for his actions and 'courage beyond the call of duty', but even though the fire service at Boscombe were at the scene of the crash very quickly, the aircraft was damaged beyond repair.

The flight test programme came to a complete halt with the loss of WD804, but despite their woes, Gloster were delighted to be told on the 7th July 1952 that the Ministry of Supply were placing a substantial production order for the GA.5 and the type was officially named the Javelin.

Even though the first prototype had been lost, the test programme was only delayed for a couple of months, as the second prototype was nearing completion. This airframe, WD808, was test flown for the first time on the 20th August 1952 and it looked very much like it predecessor. Internally though there were a number of changes and the biggest single change was the engines which were now the more powerful Sapphire ASSa6s that gave 8,200lb of static thrust. Gloster, with the announcement of the production order, were keen to show the type off to the public. It was painted with the company logo on the nose and prepared for display at the SBAC event at Farnborough during September. Because the full reasons behind the crash of WD804 had not been determined, severe restrictions were placed on WD808 and Bill Waterton was restricted to low speed, low altitude flying during the event. In spite of these restrictions the type impressed the crowds with its performance. With the event over, the aircraft returned to Moreton Valence for the continuation of the flight testing programme. The aircraft was fitted with a flight resonance system,

thus becoming the first British aircraft to carry such a piece of equipment. The system generated vibrations via a rotating eccentric cam and with it WD808 started flight trials in January 1953. The flight test schedule meant that a number of other test pilots at Glosters became involved in the process and these included Peter Lawrence (who had just joined Gloster from Blackburns), Brian Smith, Geoff Worrall and Jan Zurakowski. The whole process of testing was further speeded up with the first flight of the third prototype, WT827, on the 7th March 1953. This aircraft was built entirely at Hucclecote, unlike the previous two which had been built at Bentham and reassembled for flight at Moreton Valence. WT827 was the first airframe to be fitted with operational equipment such as the 30mm ADEN cannon and its AI radar. It was intended to undertake trials in relation to armament, although it also subsequently undertook trials for radome development, generator cooling, in-flight refuelling and general engineering. With the former the new radar

Prior to giving the first production Javelin F(AW) Mk 1, XA544, its maiden flight on 22nd July 1954, Gloster's Chief Test Pilot 'Dicky' Martin climbs aboard via a substantial set of step ladders.

system had been installed behind a Hycar radome, which was coated with Neoprene to prevent erosion. Tests soon proved that the Neoprene came off in rain or hail, so an alternative had to be found. Further tests failed to obtain better adhesion with the Neoprene, so it was decided to examine different shapes of radome to see if this solved the problem. Trials with various shapes of radome were undertaken and in the end the best option was found to be a short, bulbous one. This was fitted to WT827 when it made its first public appearance on the 15th July 1953 at the Queen's Coronation Review of the RAF at Odiham. Subsequent trials soon proved that the best shape for the radome from an aerodynamic and effective electronic point of view was decidedly more pointed, and this is the shape seen on all subsequent production aircraft. The shape of radome on production aircraft did vary slightly depending on whether British or American radar systems were installed. For the in-flight refuelling trials WT827 was fitted with a refuelling boom on the starboard wing. During flight trials this location was found to be unsuitable, the pilot found it hard to see, as it was effectively behind him, and thus made the whole operation of refuelling in the air both difficult and hazardous. To overcome this, later trials developed a new refuelling boom that was installed on the side of the fuselage and extended forward of the nose.

As early as 1952, Gloster had realised that a fairly simple modification to the Javelin's wing would increase performance and manoeuvrability at altitude. Much testing at altitude had been accomplished with WD804 and WD808 and although a sharply swept wing is good for high speed flight it does tend to increase tip stall due to the increased spanwise airflow. The modification to the Javelin's wing was achieved by adding a different angle to the leading edge sweep, just outboard of the gun ports. This 'kink' saw the inboard sections remain at 39.5°, while the outboard section was changed to 33.8°. This led to a reduction in the thickness/chord ratio and increase in the tip chord. In so doing, these changes reduced the spanwise airflow and improved the stall characteristics at the tip as well as increasing the lift co-efficiency at high subsonic speeds. A new set of wings was fitted to WD808 along with a revised and strengthened tail and a change in the elevator control ratios from 5:1 to 7.4:1. It made its first flight in this configuration on the 28th May 1953, but just two weeks later disaster struck during elevator response trials with the CofG further aft than usual. Flown by Peter Lawrence the aircraft entered a 'super stall', with the nose rising sharply as the forward speed decreased rapidly until the tailplane was completely blanketed from the airflow. Despite his best efforts he could not regain control and remaining with the aircraft for longer than he should to ensure it did not crash on a group of children playing in a field, he ejected too low and his parachute failed to deploy in time. The aircraft came down at Flax Bourton, near Bristol and when the auto-observer (the

WT827 makes a landing approach. Compared to other delta-winged aircraft, the Javelin made a much flatter approach on landing, by virtue of having a tailplane. This flat approach was laid down by the RAF as being a requirement for the all-weather fighter.
Derek N James

automatic camera and tape recorder also fitted had both failed just prior to the incident) was recovered the whole sequence of events could be unravelled. Once the stall had set in, the aircraft descended from 11,000ft in just 60 seconds although no forward speed was recorded by the auto-observer during the descent. Subsequent research showed that at speeds of less than 80kts in a high angle of attack when the flaps were deployed the aircraft entered what is termed as a 'super stall', 'stable stall' or 'deep stall'. In this situation the nose rises sharply and the speed drops off until the aircraft starts to descend almost vertically and thus the tailplane is blanketed by the wings and there is no elevator control. Much testing was undertaken by Gloster into the phenomenon and with the installation of semi-slotted flaps the nose-up pitch was reduced. The installation of leading edge slats was the best option, as this stopped the tip stall that quickly spread across the entire wing and caused the phenomena. The Air Ministry was not happy with this though, as the slats weighed 300-400lbs and they considered it an unacceptable weight penalty. In the end the problem was overcome with a simple stall warning system. This comprised a metal plate with a rotating 'mast' that was mounted on the upper surface of the wings, about mid span of the outer wing panels. When the aircraft was on the ground a spring held the plate at right angles to the line of flight. When the aircraft took off the airflow pushed the 'mast' around into the line of flight and when the undercarriage retracted it triggered a circuit to a sound box that was linked to the intercom system. Should the airflow be disturbed while in flight, the resulting drop in pressure would no longer hold the plate against the spring and the contact would close making a circuit. This circuit would thus energise the sound box and the warning buzzer would sound in the headphones of the crew to warn of the onset of a stall.

A part of the development with regard to the 'super

To coincide with the 1955 SBAC show, Gloster's Chief Photographer Russell Adams took this line-up of six production Javelin F(AW) Mk 1s, together with three of the prototypes, identified by their nose-mounted pitot head booms, while one retains its anti-spin parachute fairing atop the tailplane. XA552 and XA553 in the foreground were retained by Gloster for type equipment evaluation trials, with XA552 eventually going to De Havilland (Engines) as a Gyron Junior test-bed.

stall' also highlighted another characteristic of the Javelin and this was its behaviour in a spin. If the nose of the type was held at a high angle of attack below 100kts the airspeed would decrease rapidly until the aircraft would yaw uncontrollably to one side, then pitch to the other before spinning. The Javelin also reacted very strangely in the spin, as it was slow, but the nose could pitch up and down by as much as 70° throughout and the rotation of the spin could reverse at any moment. Much testing was done to investigate the situation, with both free-fall models and actual spinning of a Javelin being undertaken. The latter saw Wg Cdr R F 'Dicky' Martin, who had taken over from Bill Waterton in March 1954 when the latter resigned from Glosters, undertake more than 200 spins to find a way of recovery. The free-fall models were developed by the RAE and tests were undertaken with them at RAF Cardington. These trials had models dropped from a balloon at a height of 5,000ft, having been rotated prior to the drop on a turntable below the balloon to induce the 'spin' during its descent. The models were fitted with flying controls that were operated by pre-set electric motors, and a parachute system was also installed so that the model was not damaged on landing. The whole fall was filmed and various recovery procedures were tested to get the aircraft out of the spin. In the end a rather unusual system of spin recovery was developed for the Javelin. Once the stall had occurred the pilot held back the control column and centralised the rudder and ailerons, as the aircraft started to spin, into-spin aileron was applied and the

control column was pushed forward. Once the rotation stopped the control column was centralised but still held fully forward, thus the aircraft recovered either by going into an outside loop (bunt) with a moderate level of g or entering a spiral dive for which normal recovery could be used to return the aircraft to level flight.

The fourth prototype, WT830, had flown for the first time on the 14th January 1954 and this aircraft had the revised wing first seen on WD808 plus the installation of fully-powered ailerons in place of the original power-assisted ailerons seen on the previous three airframes. It was 'Dicky' Martin who, on the night of the 4th July managed to cause a sonic boom over London, in this aircraft. Subsequent comments made in the House of Lords explained that this was caused by an oxygen system failure when the aircraft was being flown at high altitude near the speed of sound, and that in the resulting confusion this had made the pilot go through the sound barrier inadvertently! This was not the first time the Javelin had gone through the sound barrier though, as it had first done so a number of weeks previously.

The fifth and last prototype, WT836 took to the air with 'Dicky' Martin at the controls on the 20th July 1954. This aircraft was more representative of the production standard Javelin Mk 1s with the kinked wing, powered ailerons, elongated radome and full operational equipment being installed. This aircraft also featured the full 'clear' perspex canopy, unlike the previous four prototypes which had all retained the partial canopy with the covered-in rear section. The upper contour of the canopy was also raised to give the crew more clearance to get hold of the grab handle of the ejection seat. Under certain conditions it had been found that the new nylon seat straps stretched and the crew found it impossible to get hold of the ejection seat firing handle. The new canopy also gave better clearance for the new 'bone dome' protective helmets that were becoming the norm for fighter pilots at this time. WT836 was used mainly for engineering trials, although it was also placed in the 'blower' rig at A&AEE for canopy jettison trials. Two days after WT836 flew for the first time, the first production F(AW) Mk 1, XA544, took off with 'Dicky' Martin at the controls.

The problem within the stall was not overcome before the type entered series production and RAF service, and on a flight on the 8th December 1955 Sqn Ldr David Dick from A Squadron, A&AEE was to experience a flat stall in F(AW) Mk 1 XA548. The flight was to determine if the production Javelin behaved in the stall and spin like the prototype and having undertaken two spins from 44,000ft he was happy that the type was very similar. As the tests continued the pilot entered a left-hand spiral at 39,000ft and while pulling 3g at 180kts the rudder slammed to the right, the spiral descent stopped and the aircraft entered a spin. Going though the usual spin recovery procedure he discovered that the aircraft did not respond as expected and the spin was flat and stable without any of the change in pitch or direction associated with the 'super stall'. All attempts were made to recover from the spin and the anti-spin parachute was deployed, but all to no avail, so the pilot ejected safely at 8,000ft. The aircraft continued to descend in the flat spin and crashed in a field near Ashley on the Isle of Wight. As a result of this crash no further spinning trials were undertaken and the stall warning system previously mentioned was developed and installed in all existing aircraft, as well as to those on the production line.

Production Tests
Of the first batch of forty Mk 1s, twenty-nine went to the RAF, but the remaining eleven were retained for trials by Gloster, Government and Ministry establishments. XA544 was used for acceptance checks of the armament and instrumentation. XA545, the second production airframe, was the first to be fitted with the 'all-flying' tailplane and XA546 was used for spinning trials and was fitted with a Gee 3 navigation system.

Both XA544 and XA546 along with prototypes WT827, WT830 and WT836 all attended the SBAC event at Farnborough in September and they all engendered considerable interest, even from a group that proved to be from the 'Russian Embassy', although they were swiftly escorted from the site and their films confiscated!

In October 1954 the news was released that following the evaluation of several European aircraft by the US Air Force, a substantial number of Javelins were to be procured. Under the Mutual Defence Aid Program (MDAP) Col R Johnson and Col P Everest had test flown a Javelin at Gloster's facility and it resulted in £36.8 million being allocated under the MDAP for the purchase of British military aircraft. All of these aircraft were to be operated by the RAF.

Sadly on the 21st October 1954 XA546 was lost in a fatal crash. The aircraft was being flown by RAF pilot Flt Lt R J Ross, who was on detachment from the RAE, when it crashed into the Bristol Channel. The subsequent enquiry determined that the pilot had tried to recover from an intentional spin at too low an altitude and the Javelin's high rate of descent gave him no opportunity to recover. Although false reports filtered in that Ross had been seen alive on Steepholme, an island in the Bristol Channel, sadly his body was never recovered.

Testing continued and XA547 was used for initial trials with the new De Havilland missile code-named 'Blue Jay'. This later became the Firestreak when adopted for service use. A long series of trials relating to spinning was undertaken by XA548, (later to be lost during a spin, as recounted previously) which was fitted with an anti-spin parachute, tail bumper, drooped wing leading-edges and slats and it was joined for a while by XA561. XA549 went to Boscombe Down for the development of radio and radar systems. In June 1955 the Javelin made its debut at the Paris Air Salon, with XA556 being displayed by Geoff Worrall. The type also saw its first service try-out when XA554 and XA559 took part in Operation Beware in early 1955. This was an annual air defence exercise and the aircraft were stationed at RAF Coltishall and crewed by Sqn Ldr P Scott, Flt Lt Jefferies, Wg Cdr E O Crew, Sqn Ldr J Walton, Wg Cdr 'Dicky' Martin and Flt Lt Williamson. The type performed well and claimed a number of Canberras as being 'destroyed' during the exercise.

On the 30th November 1955 the official CA (Controller Aircraft) Release was issued for the Javelin and the type began its service career.

On withdrawal from testing, the fifth Javelin prototype, WT836, was issued to No.2 School of Technical Training at Cosford where it remained until 1965. The photograph shows it in the early stages of scrapping, along with two Hunter F Mk 1s, in early 1965. Ray Deacon

Chapter 3: **THE 'FLAT-IRON' GOES INTO SERVICE**

XA548, the fifth production F(AW) Mk 1, was retained by Gloster Aircraft for spinning trials and is here seen with its leading-edge slats in the open position.

With the placement of the order for the Javelin by the Ministry of Supply on the 7th July 1952 the type was also given special production status. As far back as April 1952 the type had been afforded special importance when it was announced in the House of Commons by the Secretary of State for Air that the Prime Minister had 'accorded super-priority to the production of... an all weather fighter'. Even though at this stage the final decision had not been made between the GA.5 and DH.110, the end result was that when this decision was made, the winner was given 'super-priority' status. The designation of 'super-priority' had been instigated by Winston Churchill during WWII and it allowed small production runs of types at an accelerated rate if that type was considered to be of special importance to the service. This gave Gloster priority in acquiring material and manpower to meet their production schedules.

With the issuing of the CA Release on the 30th November 1955 F(AW) Mk 1s XA568, XA570 and XA572 were delivered to the RAF. These aircraft flew from Moreton Valence to No.23 MU at RAF Aldergrove on the 30th December 1955 where they underwent acceptance trials and had the AI.17 radar installed. On the 24th February 1956 the first of these was ferried to No.46 Squadron at RAF Odiham and on the 29th it was formally accepted by the RAF. The squadron, under the command of Wg Cdr F E W Birchfield DFC, were tasked with flying as many hours as possible with the type to iron out any problems that may be encountered so it could enter service with other squadrons as soon as possible. The type was a

very different beast from the aircraft previously used by the squadron, as they had flown the Gloster Meteor NF Mk 12 and 14 in the night-fighter role. The Javelin was viewed with some trepidation, as the problems with its flight characteristics coupled with the warnings about the 'super-stall' were well known. The type was also the heaviest fighter the RAF had at 31,000lb and was also the first twin-jet delta and the first all-weather interceptor that the RAF had operated. Initial flying went well, as the crews found it easy to convert from the Meteor to the Javelin but in June tragedy struck when the Commanding Officer and his navigator crashed while returning to Odiham in bad weather. The new squadron commander was Wg Cdr H E White DFC, AFC and he continued with the service testing of the type. By this time the most intensive stage was reached, whereby 1,000 flying hours were to be accumulated as quickly as possible. In an eight week period the total was reached by having aircraft take off in pairs every 30 minutes from 8.30am through to 2.30pm, then again from 6.30pm until 2.30am the next morning!

Training of crews took the form of an initial batch of pilots going to the Handling Squadron at Boscombe Down, where they were taken up in the back seat of a Javelin by a qualified pilot. During the flight the new pilot would become accustomed to the handling characteristics of the type, how it behaved in the stall and spin and just how effective the airbrakes were. These were intended for operation below 430kts, although if deployed above this speed they partially extended until speed decayed to the 430kt threshold at which point they fully deployed. The first solo flight followed fairly quickly with the pilot having his navigator with him to familiarise himself with the aircraft. It took about six weeks to convert all of the crews in the squadron to the type.

There were not too many problems with the service adoption of the Javelin, but it was very different from the Meteor and its systems did cause concern to those operating it. The type used a cartridge starter system that was both quite noisy and smokey in its physical operation, but which could also be pretty catastrophic if things went wrong. The cartridge, once fired, released cordite gas that fed a turbine starter that was engaged with the engine. This starter spun the engine up to 2,000rpm and then disengaged. However, sometimes this starter did not disengage and as the engine spun up to idle speed it was too fast for the starter and it would disintegrate! The resulting explosion could create metal shards, scarring the underside of the aircraft and the runway below it and it was a severe shock for both the crew in the aircraft and the groundcrew outside! When the starter did work, the huge plumes of dense black smoke and the odd 'whine' as the engines idled made for a very distinctive sound. The toe brakes of the Javelin were very different from the differential system on the Meteor and many a pilot went off the runway on landing by instinctively using a system they had used for so long in the Meteor! The ejection seat was distrusted by many Meteor pilots, as the NF Mk 12 and 14 did not have them fitted, and in many instances pilots refused to take the safety pins out of them prior to take-off!

During the initial stages of service acceptance with the Javelin, a special establishment was set up within No.228 Operational Conversion Unit (OCU) at Leeming. This was the Javelin Mobile Conversion Unit (JMCU), which was commanded by Sqn Ldr P D C Street DSC and comprised two Javelin pilots, four navigators, two Vickers Valleta pilots and sixteen ground crew. The role of this unit was to help all new squadrons in converting to the type and as a result it was fully mobile and saw work in both the UK and Germany. The unit would be sent to a squadron when they were due to convert and before the first Javelins arrived there. They operated two Valettas, one T Mk 4 as a flying classroom and one C Mk 1 as a transport for all their equipment and ground crew. The pilots would undergo a series of familiarisation lectures on the type, often given by Gloster staff, and then went through a very simple series of simulator sessions. These in no way reproduced the flight characteristics of the type as we would assume today, the simulator just made them familiar with the new cockpit layout and allowed them to run through pre-flight and emergency drills. The navigators undertook a number of flights in the Valletta T Mk 4, which was specially fitted with the AI.17 radar of the Javelin in an extended nose. During these simulated 'interceptions' the Valetta C Mk 1 acted as the target. After this the nav-

igators undertook thirteen flights in the back seat of a Javelin before they were considered to have been converted to type. The system worked well even though the JMCU's pilots did not have Javelins on strength to remain current, currency was maintained by collecting and delivering Javelins from various MUs. First to receive the benefit of the JMCU was Nos 23 and 141 Squadrons at RAF Horsham St. Faith during February 1957. By January 1958 the work of the JMCU was at an end, as most squadrons had converted to the type, so it returned to RAF Leeming. Here it came under the command of Sqn Ldr M 'Dusty' Miller and it did not start working again until July 1958 when the first F(AW) Mk 7s started to enter service and familiarisation flights with this more powerful version were required. A year later all the remaining Meteor units had converted to the Javelin and the unit was disbanded. It had certainly proved its worth, because in all the years it had operated not a single Javelin had been damaged during training, an enviable record that said much for the staff of the JMCU as well as the handling of the Javelin itself.

UK SERVICE LIFE

F(AW) Mk 1

During the period directly after the end of WWII the development in both aerodynamics and engines was such that new faster and higher flying designs were going into production on a regular basis. The threat from the Eastern Bloc forces was very real and the fact that their bombers were flying higher and higher made it more difficult for the RAF to actively intercept them with their current inventory of aircraft types. The main types in use during this period were the later marks of the Gloster Meteor and the De Havilland Venom. Neither could operate at the heights that the new bombers were operating at, and in many exercises the 'successful' interception was never achieved and only credited by RAF sources simply to cover up the fact that their aircraft were not up to the job. With the arrival of the Javelin all this changed, as the type was designed to undertake interceptions at 40,000ft with a service ceiling of 45,000ft. When you compare the 540kt speed at 40,000ft of the Javelin F(AW) Mk 1 with the 504kt speed at just 10,000ft of the Meteor NF Mk 12 it can be seen just how superior the Javelin was. The fact that a Javelin could reach 40,000ft in 10 minutes while the Meteor took 12 minutes to reach 30,000ft speaks for itself. This is not to say that the type was superior in all fields, its turning radius at altitude was far greater than the Venom it replaced and was also far greater than the Hawker Hunter.

The F(AW) Mk 1 equipped No.46 Squadron in the UK, thus becoming the first operational squadron with the Javelin. The F(AW) Mk 1 also equipped No.87 Squadron at RAF Brüggen in Germany and became the first overseas unit equipped with this type. These two squadrons were the only two to operate the Mk 1.

F(AW) Mk 2

The second mark, the F(AW) Mk 2, first entered service with the pioneering No.46 Squadron, in the summer of 1957. (Although the Javelin F(AW) Mk 4 preceded the F(AW) Mk 2 into squadron service by several months). The F(AW) Mk 2 was a revised version of the F(AW) Mk 4 with an American AI.22 radar installed. The squadron would retain the type when it moved from its initial base at RAF Odiham to RAF Waterbeach in July 1959 and there it continued with the F(AW) Mk 2 until it disbanded in June 1961.

No. 89 Squadron at RAF Stradishall also operated the F(AW) Mk 2, when it re-equipped from the Venom NF Mk 3 in the autumn of 1957. The squadron actually used eight F(AW) Mk 2s with eight F(AW) Mk 6s, as the latter was also starting to equip squadrons at that time. The two had contrasting performances, due to dissimilar fuel loads and therefore weights. As a result the two versions were not mixed for operations whenever that was possible. The squadron was renumbered No.85 Squadron in November 1958 and in August 1959 it moved to RAF West Malling. It remained at the base re-equipping in early 1960 with the F(AW) Mk 8.

T Mk 3

The trainer version of the Javelin originated from a design proposal (P.319) produced by Gloster to meet Operational Requirement (OR) 278 for a dual-control aircraft for pilot conversion as well as gunnery and instrument training. Initially the OR called for consideration of both tandem (fore and aft) and side-by-side seating, but to reduce severe structural redesign the tandem arrangement was adopted. The project came to naught however and the whole design was soon abandoned even though the need for a training version of the Javelin was obvious. However, in 1951 an order was placed by the Ministry of Supply for a prototype two-seat trainer and Specification T.118D was written to cover it. The aircraft was to be based on the F(AW) Mk 6 with an all-flying tail and Sapphire ASSa6 engines. The aircraft differed from the fighter by having a 44in extension in the forward fuselage. This extension had two purposes, the first being to house the second set of controls etc and the other being to counter balance the shift in the CofG (further aft) that had occurred because of the removal of the AI radar (the only radar system installed in the trainer was a small radar-ranging set). The extra space also gave the added benefit of allowing two 50 imperial gallon fuel cells to be installed, thus extending endurance. Because the type would undertake gunnery training, the cannon were retained and twin periscopes were added for the instructor in the rear seat. The instructor had a better field of view due to the larger two-part canopy sitting 10ins higher than in the fighter variants, this also gave clearance for the rear ejection seat. A single prototype was built, WT841, with the assembly being undertaken by Air Service Training using components and sub-assemblies manufactured by Gloster. Once built and ground tested the aircraft took to the air for the first time, with Jan Zurakowski at the

XD158, the prototype Javelin F(AW) Mk 2, was the first variant equipped with the American AI.22 radar, housed within a more pointed nose cone. In this photograph, the aircraft had yet to be fitted with the modified wing. Taken in early 1956, this shows Sqn Ldr Peter Scott, the RAF Liason Officer, at the controls. D James

Javelin F(AW) Mk 2, XA777, 'R' of No.46 Squadron, photographed at RAF Marham BoB display on 19th September 1959. Richard J Caruana

Close-up of the prototype Javelin T Mk 3, showing the greatly enlarged canopy and the instructor's periscope sight above the air intake.

controls, on the 26th August 1956. This aircraft was retained by Gloster for trials relating to its service entry and twenty-two production airframes were ordered in September 1954, allocated serial numbers XH390-397, XH432-438, XH443-447, XK577 and XM336. The latter two were also used for trials, with XM336 going to the CFE and XK577 going to the A&AEE. The first unit to receive the type was No.228 OCU at Leeming and throughout their service life most Javelin squadrons operated one or two T Mk 3s. (See Appendix III for details of their squadron allocations).

F(AW) Mk 4

The F(AW) Mk 4 was actually the second version to see RAF service when it started to re-equip No.141 Squadron during February 1957. The squadron was temporarily based at RAF Horsham St. Faith due to work on the runway at their normal base, RAF Coltishall. Their first aircraft, XA637, arrived on the 8th February 1957. The squadron had previously flown the Venom NF Mk 3 and they had no problems in converting to the Javelin. The Commanding Officer of No.46 Squadron, Wg Cdr White DFC, AFC was posted to command the unit so that he could pass on the experience he had gained in introducing the type into RAF service and the whole job was completed by the end of March.

The F(AW) Mk 4 had a number of improvements over the F(AW) Mk 1, the main one being the 'all moving' tailplane which reduced control column forces at high speed. The type also featured vortex generators on the wings, which increased the buffet boundary and therefore allowed the aircraft to perform better at altitude by flying closer to the stall with relative safety. The next squadron to get the type was No.23 Squadron, which was also based at RAF Horsham St. Faith. The unit used the JMCU to train their crews and received the first airframe in April 1957. In May the Javelins of all three squadrons (Nos.23, 46 & 141) took part in a large scale operation called 'Exercise Vigilant'. In it a force of 450 aircraft, including some of the new Valiant and Vulcan bombers, were used to simulate an attack on the UK and the Javelins had to intercept them way out over the sea. Although the operation showed the type up in a good light as far as operational effectiveness was concerned, there were problems with serviceability. The main problem centred around ground test equipment for the AI radar systems.

After the exercise Nos.23 and 141 Squadrons returned home and continued in their role for the night defence of eastern England. Prior to this the USAF had held responsibility for the defence of the region when operating under 'immediate readiness'. The new F(AW) Mk 4 was a far more responsive aircraft and therefore the squadrons were able to take over this role (code-named Fabulous). To meet the task two Javelins were on stand-by on special Operational Readiness Platforms to the side of the duty runway, the pilots were strapped in and the telephone scramble line was connected. There they sat, night after night, with regular crew changes every hour and a further two aircraft in readiness. In September the Javelins of both squadrons took part in the SBAC display at Farnborough and in October No.23 Squadron was detached to Nicosia, Cyprus for Exercise Dragon. No.141 Squadron used the Javelin for less than a year, as it was renumbered No.41 Squadron in January 1958.

The third squadron allocated the F(AW) Mk 4 in the UK was No.72, based at RAF Church Fenton flying the Meteor NF Mk 12 and Mk 14. It re-equipped with the Javelin from April 1959, all of the F(AW) Mk 4 airframes issued had previously been used by No.23 Squadron. To mark the event the squadron arranged a special flypast on the 10th June, Javelin XA755 flew alongside Meteor NF Mk 14 WS724 and a Gloster Gladiator K8032. The latter was flown by Gloster Chief Production Test Pilot Geoff Worrall and the formation posed a few problems considering this Gladiator was restricted to a top speed of just 184mph! During the summer of 1959 the squadron moved to RAF Leconfield and in February 1961 they moved again, this time to RAF Leeming. This was to be a temporary move while the runway at Leconfield was resurfaced, but in the end it proved to be permanent when the unit disbanded at Leeming in June.

F(AW) Mk 5

One of the main shortfalls in the Javelin so far had been operational range. To overcome this ventral tanks had been used on the previous marks, but with the F(AW) Mk 5 the situation was further improved by modifying this mark to carry an additional 125 imperial gallons of fuel in each wing. First to be equipped with the type was No.151 Squadron at RAF Leuchars. Just two airframes were officially handed over at Leuchars in May 1957, because the remaining fourteen were delivered to RAF Turnhouse while the runway at Leuchars was resurfaced. The squadron became fully operational with the F(AW) Mk 5 at Leuchars by the end of the year. The squadron remained there until September 1961, when it was disbanded.

F(AW) Mk 6

The first aircraft of this version, as has already been stated, were delivered to No.89 Squadron to operate alongside their F(AW) Mk 2s. The F(AW) Mk 6 was just a revised version of the F(AW) Mk 5, with an American AI.22 radar installed instead of the British AI.17 fitted in the F(AW) Mk 5 (much like the F(AW) Mk 2 was an 'American' version of the F(AW) Mk 4). The next unit to get the F(AW) Mk 6 was No.29 Squadron which was based at RAF Acklington. The squadron did not stay there for long though, as subsidence of the hard standing caused by previous coal mining activity, resulted in the very heavy Javelins sinking and so the unit was transferred to RAF Leuchars. In May 1960 the

Nice company in-flight image of the T Mk 3 prototype WT841. The original canopy with the raised rear seat can be clearly seen. Later the rear canopy was raised in profile above the front one.

Close up of the nose section of WT841 at an unknown location.

squadron took part in 'Exercise Yeoman', although XA829 and XA830 collided whilst in formation over the North Sea. Thankfully both crews ejected safely even though the pilot of XA830 tried in vain to get the aircraft back to a suitable airfield. The squadron continued to operate the type until the middle of 1961, when it started to re-equip with the F(AW) Mk 9.

F(AW) Mk 7

This mark was the next evolution of the Javelin design, while all previous marks had just been updated versions of the original. The type first flew on the 9th November 1956 and it featured the much more powerful (11,000lb static thrust) Sapphire Sa7 engines. It also featured a fully powered rudder and auto-stabilisers that corrected both yaw and pitch. The type also saw the introduction of the De Havilland-built infra-red guided missile that had originally been

code-named 'Blue Jay', but was called the Firestreak in RAF service. Just two 30mm ADEN cannon were retained, but due to problems in fine tuning the complex systems relating to the Firestreak many of the first F(AW) Mk 7s were actually fitted with four cannon. Without the weight of the missiles and systems and fitted with the very powerful new engines the four-gun F(AW) Mk 7s were extremely good performers. The type could climb to 45,000ft in just 6.6 minutes and its manoeuvrability at altitude was also excellent. In many ways the type outshone even the later F(AW) Mk 8s and 9s with their reheat augmented engines.

The first to operate the F(AW) Mk 7 was No.33 Squadron. This unit was, at the time on temporary detachment to RAF Leeming while the runways at its permanent base at RAF Middleton St. George was being extended and resurfaced. Because Leeming was also the home of the JMCU, when the first F(AW) Mk 7 arrived with the squadron on the 4th July 1958, the JMTU was able to quickly swing into action and convert No.33 Squadron from their previous Meteor NF Mk 12s and 14s straight onto the Javelin. Leeming was also home of the RAF Night Fighter Conversion Unit in the form of No.228 OCU and thus the radar operators of No.33 Squadron got the full benefit of their knowledge as they trained to use the AI.17 system. By mid-August the squadron had completely converted to the Javelin and by the 1st October the work at their home base was completed and they returned there. Here they took part in that year's annual air defence exercise, code-named 'Sunbeam' and the improved performance of the F(AW) Mk 7 meant that they were easily able to intercept the newer versions of the Vulcan that were now in service. The exercise did highlight the shortcomings with the ADEN cannon though, having to close to within 1,000yds of the target was an arduous task, especially at altitude as overtake speed was relatively low, (accurate use at night really required a distance of less than 50 yards). The fact was, modern Russian bombers were known to carry radar-guided rear facing cannon that could open fire from a range of 2,000yds, making the ADEN cannon obsolete in the interceptor role. To overcome this the type needed the Firestreak missiles which it had been designed to carry and the All-Weather Development Squadron (AWDS) of the Central Fighter Establishment (CFE) at RAF West Raynham was conducting trials with three missile-carrying F(AW) Mk 7s. On the 16th February 1959 No.33 Squadron sent a detachment to Nicosia, where they provided air defence for the island and undertook goodwill tours to Greece and Turkey. In April they followed this up with a visit by ten of their aircraft to Germany. On 30th September thirteen of their aircraft left Nicosia and redeployed to Akrotiri where they acted as all-weather defence for the Mediterranean region. At the end of October eight of their aircraft undertook a goodwill tour of Greece and were based at Larissa. On the 13th November the entire squadron returned to the UK, but in April 1960 they went back to Cyprus again on detachment. The squadron would eventually start to re-equip in the UK with the F(AW) Mk 9 from the 1st October 1960.

The second unit to operate the four-gun F(AW) Mk 7 was No.64 Squadron which was based at RAF Duxford. Their first machine, XH752, arrived on the 28th August 1958, followed by the second, XH747, on the 17th September, both were 'secondhand' having been previously used by the AWDS.

The next unit to get the F(AW) Mk 7 was No.25 Squadron at RAF Waterbeach. They got their first machine, XH905 and it was a fully fledged F(AW) Mk 7 complete with Firestreak missile pylons and systems.

The final squadron to get the F(AW) Mk 7 was No.23 Squadron during April 1959. At this time the squadron was temporarily based at RAF Horsham St. Faith, and it did not return to its normal base at RAF Coltishall until July that year. Once again their F(AW) Mk 7s were equipped to carry the Firestreak and this squadron was used to develop tactics for the successful operational use of this new weapon. The squadron used the F(AW) Mk 7 for just over a year and started to re-equip with the F(AW) Mk 9 in the middle of 1960.

F(AW) Mk 8

The prototype F(AW) Mk 8 made its maiden flight on 9th May 1958. The type featured two Sapphire Sa7R engines with limited reheat that increased the normal static thrust from 11,000lb to 12,300lb, but presented certain problems. The engine fuel pump fed fuel at a constant rate and if reheat was engaged at low level it reduced flow to the Sa7R combustion chamber reducing thrust. With altitude the Sa7Rs required less fuel, the 'excess' was thus available to engage reheat. As a result the reheat system was limited for use only above 20,000ft, but even then the extra thrust only just made up for the extra weight and drag of the new missile system. The reheat did improve performance in certain ways and the type could reach 40,000ft in just 7.24 minutes. The mark also featured various changes both internally and externally. The windscreen had a better rain dispersing system, a Sperry autopilot was installed and coupled to the AI radar to give automated interception. A liquid fuel starter system (AVpin) replaced the previous starter cartridge and the wings featured drooped leading edges and dampers on the pitch and yaw axis. The first operational Mk 8s (four airframes) went to the Air Fighting Development Squadron (AFDS) at the CFE on the 26th October 1959 and the first RAF squadron to receive the Mk 8 was No.41 Squadron at RAF Wattisham. Their first aircraft, XH978, arrived on the 13th November 1959. In 1960 eight aircraft from the squadron were detached to Malta operating out of Luqa airfield. Here they took part in the exercise 'Malta ADEX 60', an annual air defence exercise held on the island. Although the type performed well operationally, it had the highest unserviceability level due mainly to problems with the AI radar. In 1961 No.41 Squadron took part in the Queen's Birthday flypast over London on 10th June and in 1962 the squadron took part in the exercise 'Fawn Echo'. This was held in Norway during which the Javelins operated alongside North American F-86K Sabres of the Royal Norwegian Air Force. The squadron finally disbanded at the end of 1963 and made a special flypast over St Omer in France the location of No.41s first ever overseas headquarters in October 1916.

No.85 Squadron also used the F(AW) Mk 8 at RAF West Malling. It received its first aircraft on the 2nd March 1960. The location of this unit under a main path out of the airports in London restricted the use of the air space and as a result in September 1960 Fighter Command abandoned this base.

This No.41 Squadron T Mk 3 is seen arriving at RAF Little Rissington on the occasion of the Royal Observer Corps day, which was held on 28th August, 1960.

T Mk 3 XH436 taxies out after paying a short visit to Little Rissington in October 1961. A note on the back of the print states that it was from the Instrumentation Flight at RAF Middleton St George. Both Ray Deacon

F(AW) Mk 9

The final version of the Javelin was in fact an update of the F(AW) Mk 7 to bring it into line with the systems of the F(AW) Mk 8, no new-build F(AW) Mk 9s were ever ordered, all were converted from F(AW) 7s. Despite this converted status the type was to equip no less than eight squadrons and was the most numerous variant to see squadron service (as opposed to training or other establishment use). This mark saw the basic airframe of the F(AW) Mk 7 updated with the reheat Sapphire Sa7R engines. Other changes included the drooped leading edge to the wing and some aircraft were fitted with an in-flight refuelling boom and the capacity to carry four 230 imperial gallon drop tanks under the wings. This latter type was identified by the designation of F(AW) Mk 9R, the 'R' standing for refuelling.

The first squadron to get the type was No.25 at RAF Waterbeach. Their first aircraft were XH760 and XH767 which arrived on the 4th December 1959 and the F(AW) Mk 9 displaced the F(AW) Mk 7s flown by the squadron in the proceeding year. In September 1960 the squadron had a detachment to Cyprus and in December 1961 the unit moved to Leuchars. At this time the unit still had a detachment in Cyprus, so some aircraft flew up from Waterbeach, while others made their way across from Cyprus. The squadron was disbanded on 30th November 1962, then renumbered as No.11 Squadron the following day and transferred to RAF Geilenkirchen, Germany.

The second squadron equipped with the type was No.23 Squadron at RAF Coltishall. It got its first F(AW) Mk 9 in December 1959 and had fully converted to the type by July the next year. The squadron was the first to operate the Javelin with the in-flight refuelling system installed and many flights were undertaken by the squadron's crews to rendezvous with a Valiant tanker out over the North Sea for practice 'hook-ups'. Once the squadron was adept at the procedure it was ordered to undertake a flight to Singapore. Under the code-name 'Exercise Dyke', the aircraft were to fly out to Singapore to illustrate to the world that Britain could deploy its fighters over great distances and very quickly. Six F(AW) Mk 9s were involved and they undertook the flight in three pairs taking off on separate days. Without precise navigational aids the Javelins remained within sight of the tankers for the entire flight and the longest leg was from Gan to Singapore, when each aircraft had to be refuelled six times. This was repeated in June 1961 when the squadron took part in 'Exercise Pounce'. This saw the aircraft go to Karachi through Cyprus and Bahrain and it only took them five days on the outward leg and just four on the return home. All of this came in very handy when, no sooner had they got back than they were called on to deploy 'for real'. This time it was to Cyprus to reinforce the forces in the Middle East due to the Iraq/Kuwait crisis. The flight out had no tankers available, so the aircraft staged through France, Malta and Libya, all of the aircraft were armed with live Firestreak missiles, the first time the live operational missiles had ever been deployed by the RAF outside of the UK. The squadron remained on Cyprus for the whole of July and when ordered back to the UK on the 1st August they returned in just one day! From August 1964 the squadron started to re-equip with the Lightning.

The third RAF squadron to receive the F(AW) Mk 9 was No.64 Squadron at RAF Duxford. The squadron already operated the F(AW) Mk 7, but with the F(AW) Mk 9 and its in-flight refuelling facility, trials with the Royal Navy proved the types compatibility with the Royal Navy Scimitar and Sea Vixen aircraft when fitted with 'buddy' refuelling packs.

The next unit to re-equip was No.33 Squadron at RAF Middleton St. George during October 1960. In August 1961 the squadron was detached to Cyprus and returned to the UK in September. The squadron disbanded on 16th November 1962 and their aircraft and (some) aircrew were redeployed to RAF Germany to be reformed as No.5 Squadron at Geilenkirchen, as part of the 2nd Tactical Air Force.

The fifth and last unit in Fighter Command to operate the F(AW) Mk 9 was No.29 Squadron. This unit was based at RAF Leuchars and it received its first aircraft in the spring of 1961. The squadron had the F(AW) Mk 9 replace their F(AW) Mk 6s and as a result the radar operators had to get used to using the British AI.17, which was very different from the American AI.22 they had previously used. During 1962 the squadron deployed abroad on a number of occasions, usually going to Akrotiri. The British military authorities decided that by 1963 the all-weather defence capability in the Near East was not sufficient and therefore decided to transfer No.29 Squadron to the theatre permanently. The unit left Leuchars in February 1963 and flew to Nicosia where it took up its duties. (See Chapter 4 for overseas operations).

For Sapphire Sa6-engined Javelins, the eventual jet pipe fairing was the 'pen-nib profile shown here, on a line-up of No.141 Squadron F(AW) Mk 4s at RAF Coltishall.

XA631 was the third production F(AW) Mk 4 of the first batch built by Gloster Aircraft and was retained by the manufacturer for operational reliability trials. XA631 was subsequently released to serve with operational squadrons.

Seen at RAF Gaydon in the early 1960s is F(AW) Mk 4 XA632 'A' of No.11 Squadron from Geilenkirchen.
via Martyn Chorlton

F(AW) Mk 4, XA636 of No.41 Squadron seen here at Sculthorpe on the 17th May 1958.

Close-up of the tail marking on F(AW) Mk 4 XA636,during a visit to Sculthorpe, on the 17th May 1958.
Both Richard J Caruana

A sad sight, F(AW) Mk 5 XA646 'S', seen here on the dump at RAF Manston on the 12th June 1966. This aircraft had previously served with No.72 Squadron, 228 OCU and the AWFCS, having been struck off charge 17th September 1962 following a fire the previous July.
Richard J Caruana

In this formation of No.228 OCU Javelins, the first three aircraft are F(AW) Mk 5s, ahead of the seventh production T Mk 3, XH396, which shows the longer front fuselage. This was necessary in order to counterbalance the revised centre of gravity position created by the removal of the nose-mounted radar equipment.

By the time that the Javelin reached squadron service, a custom-designed means of entry for the crew replaced the step ladders that Dicky Martin had had to use. XH692, 'K', was the last of six F(AW) Mk 5s built by Armstrong Whitworth Aircraft, and only served with No.228 OCU during its flying career, having ended in May 1961 when it caught fire whilst landing.

Gloster test pilot Geoff Worrall lifts XA836, the twenty-second production F(AW) Mk 6 off Hucclecote's main runway. This aircraft was the last of the first F(AW) Mk 6 production batch and carried the AI.22 radar within its pointed nose cone.

Whilst XA815 was the first of the batch of twenty two mentioned in the caption above and only ever served with No's.89 and 85 Squadrons and is seen here in the latter's colours. Both were later sold as scrap on the same day, 24th June 1963.

F(AW) Mk 2 XA778 served as the prototype for the F(AW) Mk 7 series with the fitment of Sapphire Sa7 engines and is seen here in that guise. The Hawker-Siddeley company logo can be seen on the fuselage side, forward of the roundel.

Above: *The installation of two Sapphire Sa7s in the Javelin F(AW) Mk 7 brought about a further change in the rear end profile. In order to reduce a higher drag base than had originally been calculated, the rear fuselage was extended and the jet-pipe nozzles angled downwards.*

Left: *XA644 was used by Gloster in their F(AW) Mk 7 development programme. Unfortunately it was destroyed on the 24th August 1956 when it collided with a Hawker Hunter, killing the pilot Brian Smith.* Authors Collection

Below: *Early in-flight refuelling trials were conducted with the fifty-seventh F(AW) Mk 7, XH780, here operating with WH734, the Flight Refuelling Limited (FRL) Canberra B Mk 2, which was fitted with a nose probe and the necessary plumbing so that it could be used as a tanker or receiver. FRL had three Canberras which they used for the development of various hose-and-drogue units. On the Javelin, the positioning of the probe as illustrated here, was found to be OK to operate, but once the calculations for the drag it induced were made by Glosters, it was found to be excessive. As a result the large probe fitted to the starboard side of the nose/fuselage was designed and used instead.*

In the foreground, XH780 was later fitted with a veritable lance of a probe during the flight-refuelling trials. This installation, being over 20ft long, was found to be easier to use and could be fitted or removed as necessary, so that it became standard on later Javelins. Alongside XH780 is XH767, originally built as a F(AW) Mk 7 but seen here after conversion to F(AW) Mk 9 standard, with tailplane and rudder Q-feel pitot heads and cooling intakes adjacent to the fin leading edge.

A nice underside view of F(AW) Mk 7 XH712 fitted with auxiliary fuel tanks.

F(AW) Mk 7 XH714 was retained by the A&AEE for reliability trials within 'A' Squadron, but crashed in the New Forest on the 26th February 1958 killing the crew.
Gloster Aircraft Co Ltd

F(AW) Mk 7, XH899, 'P' of No.25 Squadron seen here at Sculthorpe on 16th May 1959. Was one of only 40 of its kind to be fitted with pylons for Firestreak AAMs.
Richard J Caruana

F(AW) Mk 8, XH966, was retained by Glosters for flight trials before being passed to A&AEE for high altitude armament trials. She was subsequently passed on to No.41 Squadron. Sold for scrap in November 1964.

F(AW) Mk 9 XH719 from No.60 Squadron. This aircraft remained in service until 21st August 1967 when she was written off. Both Newark Air Museum

F(AW) Mk 9R, XH889 'O' of No.23 Squadron seen during 1963.

F(AW) Mk 9F/R, XH892, 'J' of No.23 Squadron seen here during 1960. Later this airframe was allotted to the RAF Museum as Instructional Airframe No.7982M.

F(AW) Mk 9, XH904, 'T' of No.33 Squadron pictured here in June 1961.
All Godfrey Mangion

Having previously operated the earlier F(AW) Mk 7, No.25 Squadron flew the F(AW) Mk 9 for the final two years before disbanding on 18th November 1962. F(AW) Mk 9 XH880 is shown here at RAF Bentwaters in May 1960, displaying the squadron commander's individual code (Wg/Cdr J H Walton). Fred Martin

F(AW) Mk 9, XH884, 'C' of No.25 Squadron.
Richard J Caruana

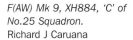

Out of the final production batch of Javelins, 40 were fitted with an in-flight refuelling probe to become the F(AW) Mk 9R variant. Displaying this enormous addition to its airframe, XH879 of No.64 Squadron from Binbrook appeared at RAF Finningley's Battle of Britain Open Day in September 1961. This squadron deployed to the Far East in 1965, operating from Tengah until they disbanded in June 1967.

Above: *Previously operated by No.25 Squadron, F(AW) Mk 9, XH880 was transferred to No.11 Squadron at Geilenkirchen West Germany after the former's disbandment in 1962. Seen at RAF Waddington's Battle of Britain Open Day in September 1963.*

Below: *No.41 Squadron replaced its earlier F(AW) Mk 4s with the F(AW) Mk 8 in 1959. XJ130 complete with a full set of Firestreak missiles poses for the camera at an*

unknown location prior to the squadron's disbandment in December 1963.

Bottom: *Although XH712 was officially an F(AW) Mk 9R, it lacked its in-flight refuelling probe when photographed here at RAF Akrotiri in the early 1960s sporting No.29 Squadron colours. During its tenure in Cyprus the squadron was deployed to Zambia during the Rhodesian crisis in 1965. All Fred Martin Collection*

Chapter 4: **THE JAVELIN ABROAD**

Although a number of UK-based Javelin squadrons had detachments abroad, there were also a number that were stationed abroad on a permanent basis. The first of these was No.87 Squadron, based at RAF Brüggen in West Germany. The unit got its first Javelin F(AW) Mk 1 in August 1957. The squadron had previously flown the Meteor NF Mk 11, so conversion to the Javelin was aided by the JMTU, which flew out from their base at Leeming. The unit marked its re-equipping and the arrival of the potential new Javelin in Germany with a special parade on the 14th August. All conversion training was completed by the end of the summer, with all fourteen Javelins in situ at Brüggen. Now operational, the squadron took part in various exercises within the 2nd Tactical Air Force region, as well as undertaking annual armament practice camps at Sylt on the West German coast. The unit remained at Brüggen as part of the 2nd TAF until it was disbanded in January 1961.

The second unit outside of the UK to receive the Javelin was No.96 Squadron at RAF Geilenkirchen, also in West Germany. The squadron got its first Javelin in December 1958 to replace their existing Meteor NF Mk 11s. The squadron was not even fully re-equipped though by the time it was renumbered as No.3 Squadron on the 21st January 1959.

As a result of the renumbering of No.96 Squadron, No.3 Squadron became the third Javelin squadron outside the UK. This squadron was also part of the 2nd TAF with two aircraft ready and armed 24 hours a day, seven days a week. The unit remained in service for two years until it was disbanded in 1961. It later reformed as part of Bomber Command, still based in RAF Germany, operating the Canberra B(I) Mk 8.

No.11 Squadron was the next unit to receive the Javelin in October 1959, although most of their F(AW) Mk 4 aircraft had previously been operated by No.41 Squadron. The squadron retained these aircraft until they were gradually replaced with the F(AW) Mk 5 from the spring of 1961, taking almost a year to complete. The squadron continued with the F(AW) Mk 5 before dispersing on 1st December 1962. However, in the UK No.25 Squadron was renumbered as No.11 Squadron on 1st December 1962, subsequently flying their F(AW) Mk 9s from Leuchars to Geilenkirchen RAF Germany.

No.5 Squadron was the next to re-equip with the Javelin, receiving their F(AW) Mk 5s from early 1960 at Laarbruch. The Berlin crisis flared up in September 1961 which resulted in the squadron being increased in strength from the usual fourteen to eighteen aircraft. Then in November 1962 (as related on page 31) No.5 Squadron re-equipped with F(AW) Mk 9s supplied by the disbanding No.33 Squadron in the UK. No.5 Squadron would now be based at Geilenkirchen to which location the F(AW) Mk 9s were flown and the 'new' No.5 Squadron was formed. By mid-December 1962 both Nos.5 and 11 would be operating their F(AW) Mk 9s from this location; their earlier F(AW) Mk 5s having been flown to the UK where most were scrapped in 1963/64.

It may seem odd to do all this swapping of squadrons, but the truth was that there needed to be examples of the latest mark of Javelin in West Germany as part of the 2nd TAF, but there were not sufficient airframes to simply re-equip them. The easiest option was just to do the re-equipping 'on paper' by moving UK-based F(AW) Mk 9 squadrons to West Ger-

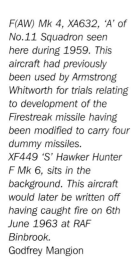

F(AW) Mk 4, XA632, 'A' of No.11 Squadron seen here during 1959. This aircraft had previously been used by Armstrong Whitworth for trials relating to development of the Firestreak missile having been modified to carry four dummy missiles.
XF449 'S' Hawker Hunter F Mk 6, sits in the background. This aircraft would later be written off having caught fire on 6th June 1963 at RAF Binbrook.
Godfrey Mangion

many by renumbering them. Both No.5 and 11 Squadrons took turns in providing the 24 hour, seven days a week quick reaction pair of aircraft to deal with airspace violations. They remained in this role for three years, before being replaced by the Lightning in 1965. The first to go was No.5 Squadron, their role being taken over by No.92 Squadron when their Lightnings arrived in November 1965. They in turn were followed on 7th January 1966 by No.11 Squadron which was replaced by No.19 Squadron and their Lightnings based at RAF Gütersloh.

Long Detachments

As has previously been mentioned, apart from units based outside the UK it was also common for UK-based Javelin units to be detached for extended periods. Two Javelin squadrons that undertook extended detachments outside of the UK were Nos.29 and 64 Squadrons. The former was transferred from their base at Leuchars to the Near East during January 1963. Based at Nicosia they gave all-weather protection in the area. The squadron quickly moved to Akrotiri where it continued in the role and in the next twelve months undertook visits for 'good will' and in support of exercises to Aden, Gibraltar, Iran, Malta, North Africa and Singapore. The limited range of the Javelin was a problem considering the vast areas that the squadron was operating over and this was overcome by re-equipping with the F(AW) Mk 9R. This was simply a standard F(AW) Mk 9 that had the capability to be fitted with an in-flight refuelling probe and to carry four 230 imperial gallon tanks on underwing pylons. The refuelling probe was not a permanent fixture and was usually only used for transit flights. Although four fuel tanks could be carried it was usual for just two to be fitted, as the weight penalty of all four as they filled was such that the Javelin had to adopt a higher angle of attack to keep station with the tanker and use reheat, thus depleting the fuel almost as quickly as it was being taken onboard! During such long deployments the Javelins needed to receive fuel on a large number of occasions per flight, simply so they had enough fuel on board at any time to divert to a suitable air base should a problem arise.

In 1965 No.29 Squadron was involved in the detachment in support of the defence force to Zambia. The Unilateral Declaration of Independence by Ian Smith in Southern Rhodesia at this time had made the British Government fear that this 'new' nation would launch attacks on the hydro-electric plants at Kariba in nearby Zambia. The British Government offered Zambia a defence package, which was accepted, and a large part of this force was made up of ten F(AW) Mk 9Rs from No.29 Squadron. They departed from Akrotiri on the 1st December 1965 and flew non-stop to Nairobi in Kenya, making an illegal incursion into Egyptian airspace as they went, much to the disgust of President Nasser! They then flew from Nairobi to the civil airport at Ndola in Zambia, where

they received the additional protection of a detachment of troops from the RAF Regiment. Four of the aircraft were relocated 200 miles south at Lusaka, but neither location had any form of modern facilities and all accommodation etc was very makeshift. It soon became obvious that the whole situation was not as potentially dangerous as had been made out by the government and the Rhodesians had no intention of attacking the dams in Zambia. Just six months later the decision was made to withdraw the Javelins from Ndola and Lusaka and the Javelins departed to return to Akrotiri at the end of July 1966. This time their return flight was routed via Kenya and Aden to avoid illegally overflying Egyptian airspace. Back at Akrotiri the squadron returned to its usual tasks with visits to Malta for the Annual Practice Camps but their time in the Middle East came to an end in April 1967 when their role was taken over by the Lightnings of No.56 Squadron. No.29 Squadron flew back to the UK and relinquished their Javelins for Lightnings on the 1st May 1967.

No.64 Squadron's extended period outside the UK followed a number of short detachments to the Middle East. They were then re-equipped with the F(AW) Mk 9R and selected for 'long-range' detachments in support of forces in the Middle and Far East. The first such detachment happened in 1963, when the squadron sent twelve of their aircraft to India. Here they took part in 'Exercise Shiksha'; this exercise was

F(AW) Mk 7, XH719, 'J' of No.33 Squadron seen here during an Open Day at Ta' Qali, Malta (date unknown).

The last F(AW) Mk 7 manufactured by Armstrong Whitworth, XH899 seen after conversion to F(AW) Mk 9R at Luqa, Malta on 27th September 1965. Both R J Caruana

due to the Nassau Agreement that pledged support by the US and UK for India should that nation be threatened by invasion. At this time India was under increasing threat caused by a number of incursions by Chinese troops along their Northern Frontier. No.64 Squadron left their base at RAF Binbrook on the 27th October and their arrival the next day marked the first time that British warplanes had entered India since it had gained its independence. The initial role of the Javelins during the exercise was to act as the attacking force, although later in the exercise they reverted to their true role as interceptors. The exercise was also supported by F-100s from the USA, Canberras of the RAAF and numerous Canberras, Gnats and Hunters of the Indian Air Force. By the time the exercise came to

an end just seven Javelins went back to Binbrook. One had crashed during the exercise with an American exchange pilot at the controls, and four others were onward ferried to Singapore where they joined No.60 Squadron as reinforcements (see page 45 Far East). Back at Binbrook the squadron took on its usual home defence role, but the situation in the Far East was worsening, so the decision was made to send a Flight of eight aircraft to Singapore. The flights out took a considerable amount of organisation and used both landing stages and long legs with in-flight refuelling to get the Javelins to Tengah, but they all arrived there during the period from the 11th to the 21st September 1964. Back in the UK the remaining aircraft with the squadron were gradually depleted, with most going to No.29 Squadron at Akrotiri. The Flight at Tengah was increased and on the 1st April 1965 it took over the mantle of the squadron so that No.64 Squadron was now based entirely in the Far East. The unit now took on the role of support in relation to the escalating aggression towards Malaya that was being shown by Indonesia in what was referred to as the 'Confrontation Crisis'. This saw the squadron fully armed and detached to Labuan and Kuching in Borneo, a job that had previously been done by No.60 Squadron. The unit continued in this role until June 1967 when it was officially disbanded.

A rather well worn F(AW) Mk 9 XH892 of No.29 Squadron seen here at Luqa on Malta, believed to be 6th April 1967. Richard J Caruana

F(AW) Mk 9R, XH895, 'P' of No.23 Squadron. This aircraft was later delivered to No.60 Squadron after serving with No.64 Squadron. Godfrey Mangion

These Javelins of No.64 Squadron passed through Malta on delivery to No.29 Squadron on 20th February 1964. 'F' is an F(AW) Mk 9R, carrying the serial XH762. R J Caruana

Far East

No.60 Squadron was based at RAF Tengah Singapore and converted to the Meteor NF Mk 14 in November 1959, thus becoming the first night and all-weather fighter squadron within the Far East Air Force (FEAF). The unit was to re-equip with sixteen Javelin F(AW) Mk 9s in July 1961 and these were ferried out in batches with a 20 hour flight via ten countries and fifteen staging posts! The first batch arrived safely on the 8th August, but only four of the five aircraft in the second batch arrived. Unfortunately one of the aircraft, XH791, suffered a double engine failure over East Pakistan (now Bangladesh) and even though both crew ejected, only one survived. In the third delivery batch just three made it initially, as one was damaged at Tehran and took a considerable time to be repaired and another one was damaged beyond repair in a refuelling incident at Luqa, Malta.

The involvement of No.60 Squadron in the troubles in Malaya first started in March 1962, when they had a 'presence' based at Butterworth in Northern Malaya. The situation in the area had been tense for a considerable time, but it was not until April 1963 that Indonesia started to actively confront the new Federation of Malaysia, comprising Malaya, Singapore, Sarawak and Sabah. The first incursions into Eastern Malaysia (North Borneo) from Indonesian Borneo took place at this time and in September the 'Confrontation Crisis' was officially acknowledged. No.60 Squadron was tasked with support of the federation and protection of its borders, however this resulted in the Javelin being used to fly long, low-level border patrols and this was something that the type had never been designed to do. The missions were often flown as low as 50ft, over dense jungle in appalling weather with no navigational aids. Low-level flying in such humid conditions did not suit the Sapphire, although the pilots often flew on just one engine with the other idling to save fuel. The Sapphire was optimised for use at high altitude, so the strains put on it in the low-level role resulted in compressor blade failure and to overcome this restrictions were imposed to prevent the use of the engine in the 82-93% rpm range, much to the annoyance of the pilots. Another serious problem was 'centreline closure', which occurred when super-cooled water in cumulo-nimbus clouds was ingested into the engine and caused the compressor casing to contract until it was in contact with the compressor blades. The result was catastrophic and the aircraft lost on the second delivery flight to No.60 Squadron suffered this phenomenon. To overcome this, in true British style, abrasive pads were added to the inside of the compressor casing so that when the casing shrank, it just wore the compressor blades away! No.60 Squadron had its first operational loss in the 'Confrontation Crisis', when XH836 suffered hydraulic failure while on patrol and both crew had to eject. They both survived, but it took rescue forces five days to find them.

In October 1963 it was decided to counter Indonesian incursions into Malaysian airspace over Penang with a detachment of Javelins based at Butterworth, which was used by the RAAF. The two Javelins initially sent proved insufficient in quantity so on the 10th December they were joined by four more supplied by No.64 Squadron which had come from India having participated in 'Exercise Shikska'. In January 1964 this six-aircraft detachment was made permanent and designated as C Flight. At the same time the Indonesian Government made it clear that they intended to use aircraft to support their troops making incursions into Malaya, so an Air Defence Identification Zone (ADIZ) was declared on the 26th February and eight Javelins from Tengah were flown out along with Hunters of No.20 Squadron to Labuan and Kuching to patrol the ADIZ. The extra pressure now imposed on No.60 Squadron due to the need to keep two aircraft on constant armed stand-by resulted in them being reinforced with four more F(AW) Mk 9Rs from store at No.27 MU Shawbury in the UK. To mislead the Indonesians, these aircraft arrived bearing No.23 Squadron markings to create the impression that they were on a short-term detachment to Tengah, but they were to be permanent residents there joining B Flight. By this stage the squadron had become the biggest in the RAF with twenty-six Javelins on strength, it was to

F(AW) Mk 9, XH793, 'A' of No.23 Squadron at El Adem, 11th January 1963. Richard J Caruana

get bigger though and had a peak strength of thirty-three aircraft and forty-two operational crews. Although the 'Confrontation Crisis' had reached a peak in 1965 the feared all-out war did not occur and in 1966 the Indonesian forces reduced their activities and the crisis was officially ended on the 11th August. Prior to this, during April, the strength of No.60 Squadron had reduced to just twenty-two aircraft, with twelve at Tengah and ten at Butterworth. By June 1967, with the disbandment of the only other Javelin squadron in the area (No.64), No.60 was on its own. During the month it had to deploy four Javelins to Kai Tak Airport in Hong Kong due to riots caused by Communist factions, but they only remained there for nine days before being withdrawn. By April 1968 the unit had reduced to just ten aircraft all based at Tengah where preparations were being made to hand over the role to Lightnings. The disbandment ceremony took place on the 30th April and so ended the service career of the Javelin with the RAF.

Javelins belonging to No.29 Squadron bask in the sun in Yard One, RAF Luqa, Malta in 1966. The nearest aircraft is an F(AW) Mk 9 XH834 'Z' while next to it is T Mk 3 XH396 'D'.
Richard J Caruana

No.23 Squadron F(AW) Mk 9R, XH849 'C' taxies in after flying non-stop to Aden in late 1962. In-flight refuelling was via two Valiant tankers XD816 and XD817.

No.23 Squadron F(AW) MK 9Rs XH889 'L', XH849 'C' and XH890 'M' seen lounging in the sun at Khormaksar after their long flight from the UK in late 1962. The sun screens over the canopy look like beds, but they are in fact official 'sun shields' and an illustration of this appears elsewhere in this title. Both Ray Deacon

XH889 and XH890 of No.23 Squadron in the company of their Valiant tankers XD816 and XD817 at Khormaksar in late 1962 after their non-stop flight from the UK.

F(AW) Mk 9R XH890 of No.23 Squadron on arriving at Khormaksar in late 1962.

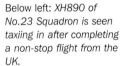

Below left: XH890 of No.23 Squadron is seen taxiing in after completing a non-stop flight from the UK.

Above right: F(AW) Mk 9R, XH766 'E' of No.64 Squadron after arriving at Khormaksar for an overnight stay in 1963. These machines were probably on route to Tengah to reinforce No.60 Squadron, there during the unrest in the Far East at this time. (See also page 110).

Left: F(AW) Mk 9R, XH893, 'V' after arrival at Khormaksar on route to Tengah in 1963 to reinforce No.60 Squadron. All Ray Deacon

Chapter 5: JAVELIN TECHNICAL DESCRIPTION AND ARMAMENT

Fuselage

This was built in four main sections, two of which were permanently joined during the manufacturing process. This was also the biggest section and comprised the front and centre sections. It was made up of a main spar frame, centre structure and six frame-to-skin stringer sections. The main spar frame was the most important element, being load-bearing and having fork-end couplings at either end for attachment to the main wing spars. In the centre of this beam there was considerable reinforcement and attached to it at this point was the keel member (or box beam centre structure) that ran the entire length of the centre section. This keel member had the vertical fin and rear wing attachment points on it. Sections made up of skins attached to longitudinal stringers created the rear fuselage section. There were three main sections, attached to the keel member to form the engine bay. The lower sections also had runners on which the engine could be installed. The final sections of the centre section were the engine air intakes. These were made as separate structures and attached directly to the fuselage sides, which housed the air intake ducts coupled directly to the engine compressor casing.

The front fuselage section contained the nose wheel well and cockpit. The latter was pressurised and sealed at both the front and rear. The canopy featured an armoured windscreen and all ancillary equipment associated with the electrically-operated sliding canopy. Some element of ballistic protection was afforded the crew by the use of heavy gauge aluminium in the cockpit flooring and sidewalls. A large bay was situated below the rear cockpit station and this contained the engine starting equipment, auxiliary gearbox, hydraulic panel, generator control, electrical distribution panel and some of the radio equipment. The bay could be accessed via a large removable panel in the lower skin of the fuselage directly below it.

The rear fuselage section was of light construction in semi-monocoque (continual, semi-rigid) form. It incorporated the engine exhaust outlets which were attached to the centre section with quick-release fasteners.

The radar was enclosed in a Hycar radome and on those marks that had American radar equipment installed (F(AW) Mk 2, 6 & 8) this unit was larger and was fitted on hinges that allowed it to be pulled forward and swung to one side for access to the radar unit.

The fin featured two spars and was constructed in the usual manner with leading edge and inter-spar ribs. Attachment points in the form of plates and Z-shaped pick-ups were situated at the top of each

spar for the fixture of the small non-moveable section of the tailplane.

The tailplane had a straight tubular spar at its centre that was about 1/3rd the total width of the entire tailplane. It was extended to its full width by box spar extensions that were swept back. The whole assembly was covered in aluminium skins and the rear edge of the torsion box was formed by the curved fairings for the elevators.

Wings

The delta design was rigid and therefore the actual structure of the Javelin wings was surprisingly simple. The wing attached via three points and the necessary rigid 'torsion box' elements were formed by the main spars and the leading edge ribs and their skinning. The inboard section was a separately built sub-assembly that housed the fuel tanks, undercarriage, flaps, airbrakes and armament. During manufacture it was joined to the trailing edges which contained the Servodyne control units for the ailerons and the mass-balanced ailerons themselves. The outboard panels were complete sub-assembles that were attached to the inboard sections with two fish-plates. These took the full range of compression and torsion loads that would be exerted on them in flight. Once assembled rivetted joints were used to complete the torsion box and this took some degree of shear loading.

The primary join for the wing to fuselage was the root rib. This join transferred the bending stresses across the fuselage through the main spar. As a crucial load-bearing element it should be noted that the assembly failed at 114% of ultimate design load on the prototype and 115% on the much heavier production airframe.

Unlike other deltas, because the Javelin had a separate tailplane, the type had conventional flight controls. The wing therefore had the usual ailerons and flaps, while the tail had elevators and a rudder. The first three aircraft had power-assisted flight controls powered by hydraulic units at a ratio of 5:1 for the ailerons and elevators and 7.4:1 for the rudder. To overcome the lack of 'feel' power controls give through the control column a degree of aerodynamic forces exerted on the control surfaces was transferred back to the control column. Fully-powered ailerons with a hydraulic Q-feel simulator to give 'feel' through the control column were installed from the fourth aircraft. Large slotted airbrakes were situated in the upper and lower inboard wing panels, inboard of the trailing edge. These were intended for operation below 430kts, although if deployed above this speed they partially extended until speed decayed to the 430kt threshold when the airbrakes then fully deployed. In the Javelin these airbrakes were rapid to deploy and extremely effective.

Tail

The tailplane for all of the prototypes, most of the F(AW) Mk 1s and some F(AW) Mk 2s was electrically-operated by the trim switch on the control column. Later marks had a hydraulically-operated 'all-flying' unit that had the electric motors of the previous version replaced with a hydraulically-driven screwjack. This system was directly operated from the control column. The elevators thus acted as anti-balance tabs with proportional load in relation to airspeed induced by twin simulators. These tabs automatically moved in the opposite direction to the tailplane. Using the switch on the control column the pilot could trim at residual stick forces following a change in flight conditions. eg. increase or decrease in speed or the lowering of the undercarriage.

Engines

All marks of the Javelin were powered by versions of the Armstrong-Siddeley Sapphire. The marks up to and including the Sa6 were powered by versions of the Series 100 Sapphire, while those after it had the Series 200 installed. The Sapphire had a thirteen-stage axial flow compressor, a single annular-type combustion chamber with two high-energy ignition plugs and a two-stage axial turbine. The jet exhausts were fixed in all versions except for those fitted with reheat (Mks 8 and 9), which had a variable exhaust nozzle system. This nozzle utilised twenty-seven inter-leaved segments and was hydraulically operated.

The breakdown of Sapphire usage is shown below.

F(AW) Mk 1

Armstrong-Siddeley Sapphire Sa6 Mk 10201 (port) and 10301 (starboard) turbojets rated at 8,300lb static thrust.

F(AW) Mk 2

Armstrong-Siddeley Sapphire Sa6 Mk 10701 (port) and 10801 (starboard) turbojets rated at 8,300lb static thrust.

T Mk 3

Armstrong-Siddeley Sapphire Sa6 Mk 11201 (port) and 11301 (starboard) turbojets rated at 8,000lb static thrust.

F(AW) Mk 4

Armstrong-Siddeley Sapphire Sa6 Mk 10201 (port) and 10301 (starboard) turbojets rated at 8,300lb static thrust.

F(AW) Mk 5

Armstrong-Siddeley Sapphire Sa6 Mk 10201 (port) and 10301 (starboard) turbojets rated at 8,300lb static thrust or Armstrong-Siddeley Sapphire Sa6 Mk 10701 (port) and 10801 (starboard) turbojets rated at 8,300lb static thrust.

F(AW) Mk 6

Armstrong-Siddeley Sapphire Sa6 Mk 10701 (port) and 10801 (starboard) turbojets rated at 8,300lb static thrust.

F(AW) Mk 7

Armstrong-Siddeley Sapphire Sa7 Mk 20301 (port) and 20401 (starboard) turbojets rated at 11,000lb static thrust.

F(AW) Mk 8

Armstrong-Siddeley Sapphire Sa7R Mk 20501R (port) and 20601R (starboard) turbojets rated at 11,000lb static thrust and with 12% reheat to give 12,300lb static thrust above 20,000ft.

F(AW) Mk 9 & 9R

Armstrong-Siddeley Sapphire Sa7R Mk 20901R (port) and 21101R (starboard) turbojets rated at 11,000lb static thrust and with 12% reheat to give 12,300lb static thrust above 20,000ft.

AUXILIARY SYSTEMS

Hydraulics

The hydraulics in the Javelin powered the flying controls, undercarriage, airbrakes, wheel brakes, nose wheel shimmy damper and oleo leg charging system. Three internal pumps were driven by an auxiliary gearbox running off either engine and it provided 3,000lb/sq. in. pressure. This system meant that even with an engine failure it did not result in the loss of any of the auxiliary systems. One pump drove the system for the undercarriage, flaps and airbrakes, while the other two were duplicate systems for the flying control. Hydraulic fluid was contained in two tanks so that should one of the tanks fail, it would not cause a complete failure. Smooth pressure was assured by three hydraulic accumulators built into the system.

The nose and main undercarriage had Dowty liquid-spring shock absorbers fitted and these were recharged in flight via the main hydraulic system. Maxaret anti-skid units were fitted to each of the Dunlop disc brakes. The nose wheel could castor and was fitted with a shimmy-dampened, self-centering system. It had no brakes fitted, so the wheel was stopped from spinning in the wheel well by a spring-loaded damper with a pad on it that contacted with the wheel and stopped it once it was fully retracted.

Electrics

Electrical power was produced by two 6,000W generators as a DC current. This supplied the whole electrical system and also charged the on-board battery. A converter to change the current to AC was installed for supply of the flight instruments, navigational and radio equipment and radar system.

Engine Starting

This was one of the least successful systems in the Javelin. On the F(AW) Mks 1 to 6 the system installed was a cartridge starter. This used two 5in starter cartridges, that once fired, the expanded cordite gas was fed to a turbine starter. This starter spun the engine at 2,000rpm, at which point it was to disengage. Unfortunately this disengagement did not always happen and as the engine spooled up, the starter unit would explode due to overspeeding and thus cause considerable damage to the airframe. The system also suffered many failures of the feed line from the cartridge to the starter unit, failure of the starter unit, non-firing of the starter cartridge or the simultaneous firing of both primary and back-up cartridges.

In the F(AW) Mk 7, 8 and 9 the cartridge system was replaced with a liquid starter. This used a small cordite charge to ignite the liquid fuel (AVpin) in a reaction chamber. The gas created increased the pressure in the chamber and thus drove the starter unit. Three cartridges were installed, although often they failed to fire. If another set was installed and failed, the whole system had to be left for an hour before the procedure could be repeated. The cartridges were situated in a bay aft of the rear cockpit and the cover over it was usually left open when starting, as a misfire could cause damage to the hatch or, at worst, a fire.

Armament

Many armament systems were considered for the Javelin and these are discussed in Chapters 1 and 6, so I will restrict my coverage here to the armament systems actually fitted to the Javelin during its service life.

ADEN Cannon

The 30mm ADEN cannon was developed from the 20mm MG 213C which had been produced in Germany during WWII. Initial work had been started by Mauser in 1943, but by the time the war ended just five versions had been completed and the weapon was captured by the Allied forces. The MG 213C was a belt-feed, electrically-fired, pneumatically-charged weapon and its development post-war in the UK was instigated by the Ministry of Supply. This development took place at the Armament Development Establishment at Enfield, which is why it is called ADEN (**A**rmament **D**evelopment **E**stablishment at E**N**field). The original 20mm calibre was increased to 30mm and by reversing certain components within the weapon it could easily be changed from left to right-hand feed. The gun was a rotary unit, with five chambers that rotated so that each chamber could be brought in line with the barrel when in the 12 o'clock position. The weapon could fire 1,200rpm and could withstand 150 rounds being fired in a single burst without it overheating.

The installation of the ADEN in the Javelin was straightforward and each wing contained two guns with their associated ammunition and feeds. Each set of two guns was heated via air bleed off the area adjacent to the engine compressor casing. The gun bays were purged of cordite gases when the weapon discharged by the constant flow of air into the bays. The first Javelin to actually carry the weapon was WT827 and initial trials showed that airflow was passing into the barrels and heating the ammunition. Frangible covers were tested, but these overheated and melted as within 10 minutes the temperature of the muzzles had reached 124ºC. A number of tests with barrels of various lengths eventually solved the problem, the best option being a muzzle that protruded from the wing by 0.25in.

Firestreak

The ADEN cannon fitted to all marks up to the F(AW) Mk 6 was always considered an interim measure, as the type was intended to carry air-to-air missiles. De Havilland developed an infra-red guided air-to-air missile under the code-name 'Blue Jay'. This missile was later accepted for service use and was called the Firestreak. Unlike modern missiles the Firestreak could not 'see' its target unless it was directly behind it. During the initial stage of the interception the AI radar locked on to the target and the navigator would direct the missile head where to look. The Firestreak guidance system would lock on when it was in a conical area behind the target and the pilot would not launch it until he was advised he was in range by the navigator. The fitment of the missiles to the Javelin was a complex procedure, even the pylons on which they were carried were an aerodynamic marvel. The pylons were fitted in pairs beneath each wing, and were swept back to the wing when viewed from the side. From the front the thickness-to-chord ratio was 10% at the top and 6% over the missile. On top of this the maximum thickness chordwise moved back progressively from the bottom to the top (Kuchemann 'waisting'). All of this meant that even with four missiles in place the effect on performance and range was negligible. The Firestreak was powered by a solid fuel rocket motor with a burn time of three seconds. The maximum range was four miles, although it could be launched from as close as one mile. The missile was able to climb 10,000ft from launch altitude and the 50lb warhead was was triggered by a proximity or contact fuse.

The first aircraft to carry the missile was F(AW) Mk 4 XA632 in July 1956 and a long series of flight trials was instigated before it was adopted for service use. The main trials organisation was No.1 Guided Weapons Test Squadron at RAF Valley. This unit had previously been the Guided Weapons Development Squadron and received its first Javelin (XH701) on the 13th January 1959. Trials ran from January 1959 through to May 1962 during which ninety-nine Firestreaks were launched. The first Javelin to carry the

Firestreak operationally was the F(AW) Mk 7, although of the 142 Mk 7s built, just 42 of them were equipped to carry the missiles. The remaining 100 aircraft carried no missiles and retained all four cannon instead. The F(AW) Mk 8 and 9 were both equipped to carry missiles and all of these marks retained one ADEN cannon in each wing when carrying the Firestreak.

The Firestreak ceased production in 1960, but the missile was still being used with the Lightning as late as 1983.

One of the finest examples of the Javelin to survive to the present day is F(AW) Mk 8, XH992 at the Newark Air Museum, Winthorpe, Lincolnshire. Resplendent in No.85 Squadron's colours, it is displayed with Firestreak missiles. Fred Martin

This cutaway from the manual shows the two-seat T Mk 3.
© Crown Copyright

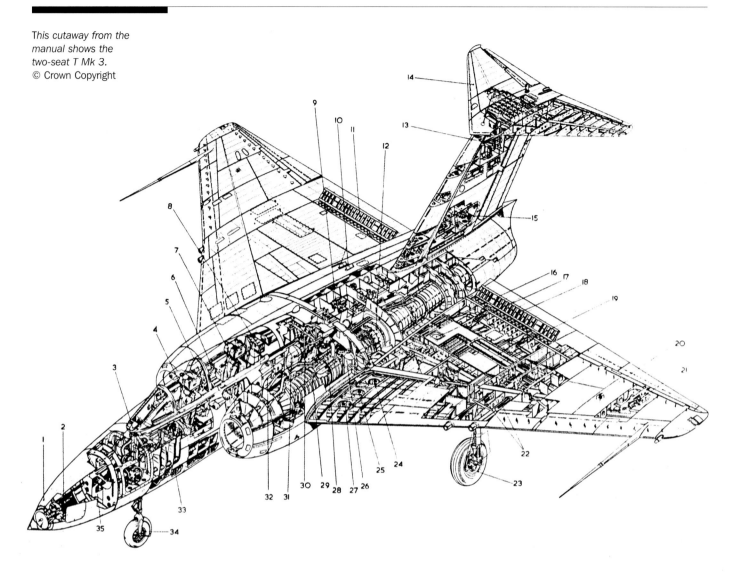

Key
1 Radome
2 Radar head
3 Gyro gun sight
4 Pupil's ejection seat
5 Periscope
6 Gyro gun sight
7 Instructor's ejection seat
8 Zero reader computer
9 Auto pilot servo motors
10 Feel simulator pressure heads
11 Top airbrake
12 Sapphire gas turbine
13 Tail plane power control unit
14 Movable tail plane
15 Rudder servodyne
16 No 5 wing fuel tank
17 No 4 wing fuel tank
18 Landing flap
19 Ammunition tanks
20 Stall warning vane
21 Aileron servodyne
22 30mm ADEN guns
23 Mainwheel unit
24 No 3 wing fuel tank

25 No 2 wing fuel tank
26 Fuselage side tank
27 No 1 wing fuel tank
28 Cabin primary cooler
29 Air intake duct
30 Accessories gearbox
31 Hydraulic reservoir
32 Fuselage centre tanks
33 Oxygen bottles
34 Nose wheel unit
35 Oxygen bottles

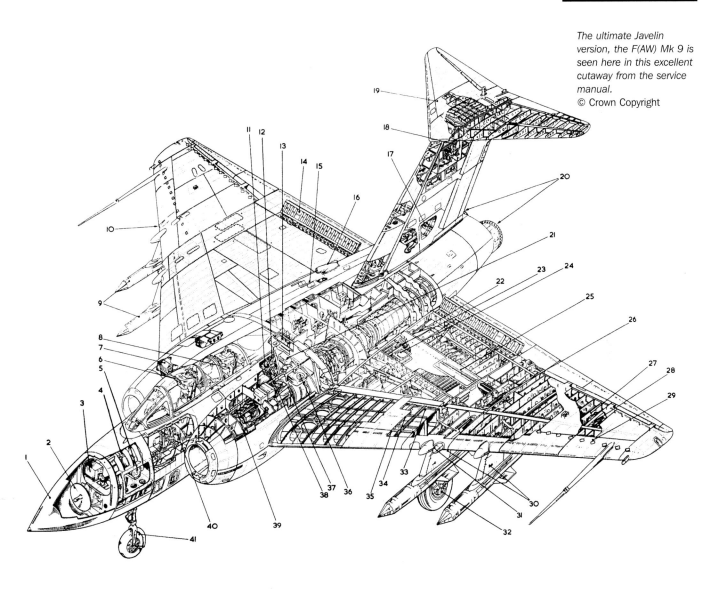

The ultimate Javelin version, the F(AW) Mk 9 is seen here in this excellent cutaway from the service manual.
© Crown Copyright

Key

1 Dielectric nose
2 Scanning unit
3 Transmitter/receiver
4 Amplifier
5 Modulator
6 Pilot's ejector seat
7 Mark Space amplifier
8 Navigator's ejection seat
9 Firestreak missiles
10 Cine camera
11 Hydraulic pumps
12 Accessories gearbox
13 Tail plane simulator
14 Top air brake
15 Simulator pressure head
16 Zone 3 cooling scoop
17 Rudder servodyne
18 Tail plane power control unit
19 Movable tail plane
20 Reheat jet pipe nozzle
21 Cabin primary cooler
22 Landing flap

23 No 5 wing fuel tank
24 No 4 wing fuel tank
25 Ammunition tanks
26 30mm ADEN guns
27 Pitch actuator and potentiometer
28 Aileron servodyne
29 Vortex generators
30 Missile pylons
31 Blast tube fairing
32 Main wheel unit
33 No 3 wing fuel tank
34 No 2 wing fuel tank
35 No 1 wing fuel tank
36 IPN fuel tank
37 Sapphire gas turbine
38 Hydraulic alternator
39 Air intake duct
40 Pilot's instrument panel
41 Nose wheel unit

This drawing of the F(AW) Mk 9 shows the in-flight refuelling probe and other detail changes incorporated in this version. © Crown Copyright

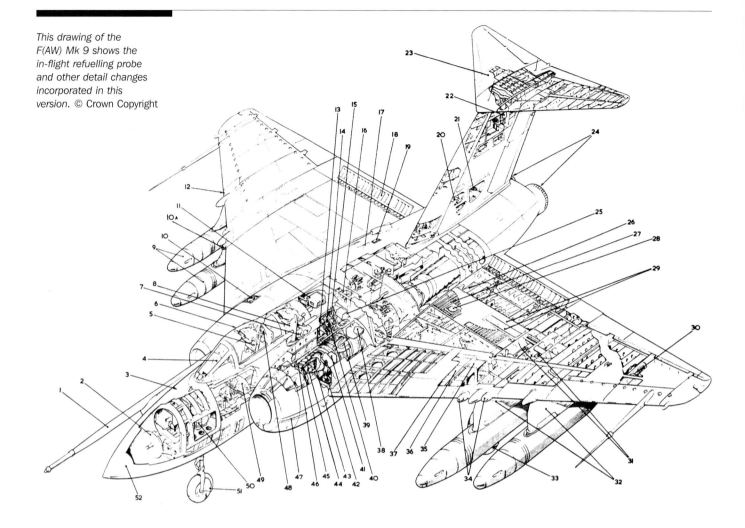

Key

1 In flight refuelling probe
2 Scanner
3 A I bay
4 Gyro gunsight
5 Modulator
6 Pilot's ejection seat
7 A D F receiver
8 Navigator's ejection seat
9 Under wing pylon tanks
10 VHF/UHF aerial stowage
10A Refuelling lamp
11 ADF loop aerial
12 Cine camera
13 VHF sets 1 & 2
14 Accessories gearbox
15 Generators
16 Tail plane simulator
17 Simulator pressure head
18 Top air brake
19 JPT test plug
20 Fire wire control unit
21 Rudder servodyne
22 Tail plane power control unit
23 Movable tail plane
24 Reheat jet pipe nozzle
25 Cabin primary cooler
26 Landing flap

27 No 5 wing fuel tank
28 No 4 wing fuel tank
29 Ammunition tanks
30 Aileron servodyne
31 30mm ADEN guns
32 Under wing tank pylons
33 Main wheel unit
34 Blast tube fairings
35 No 3 wing fuel tank
36 No 2 wing fuel tank
37 No 1 wing fuel tank
38 IPN fuel tank
39 Sapphire gas turbine
40 Hydraulic alternator
41 Armament power pack
42 Rebecca T R set
43 UHF junction box
44 UHF transmitter/receiver
45 VP junction box
46 Air intake duct
47 Rebecca homing aerial
48 ADF controller bearing indicator
   and Rebecca range and heading meter
49 Pilot's instrument panel
50 Additional oxygen bottles
51 Nose wheel unit
52 Dielectric nose

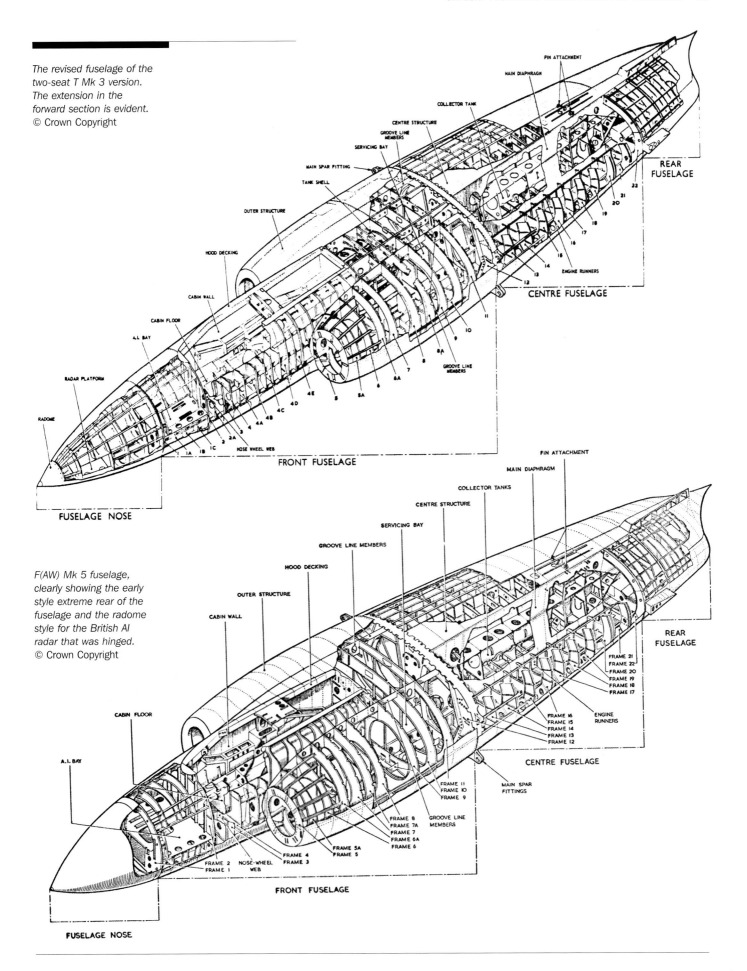

The revised fuselage of the two-seat T Mk 3 version. The extension in the forward section is evident.
© Crown Copyright

F(AW) Mk 5 fuselage, clearly showing the early style extreme rear of the fuselage and the radome style for the British AI radar that was hinged.
© Crown Copyright

This truncated view of the extreme front section of the F(AW) Mk 8 shows the hinged radome fitted to all versions with American AI radar. Note that the hinge line is not at 90 degrees to the centre line.
© Crown Copyright

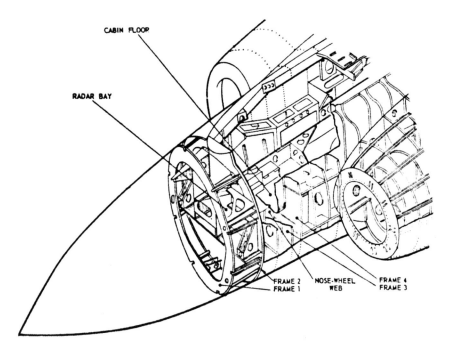

This drawing from the service manual shows the construction of the fuselage for the early marks without the tail cone extension.
© Crown Copyright

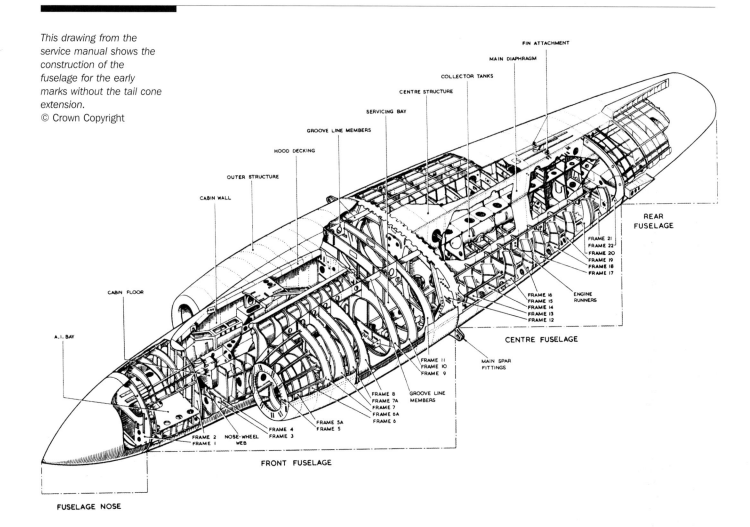

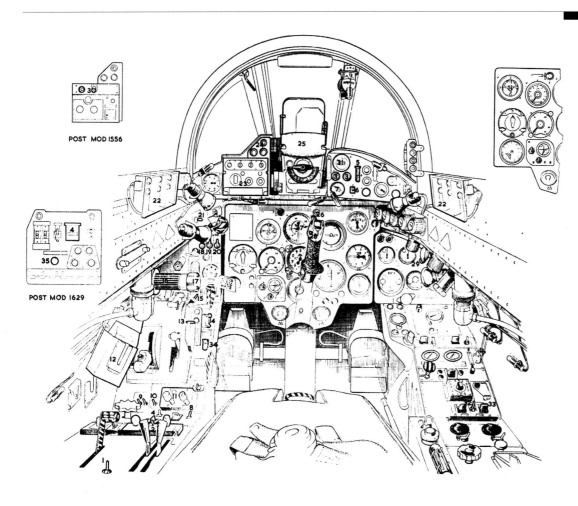

This diagram from the flight manual shows the front cockpit of the F(AW) Mk 9.
© Crown Copyright

Key

1 Guided weapons re-set switch
2 ARI 5807 controller
3 Guided weapons arming indicator lamp
4 Guided weapons or rocket battery selective jettison switch
5 Guided weapons arming switch
6 Guided weapons pairs/single switch
7 Master armament selector switch
8 Rocket battery selector switch
9 Gun scavenging emergency switches
10 GGS on/off switch
11 GGS control unit
12 GGS recorder stowage
13 Guided weapons, rocket battery or pylon drop tank jettison switch
14 Emergency intercom and tele-briefing call lamp and push switch
15 UHF aerial change-over switch
16 GGS range control twist grip
17 P/T and P/M switches
18, 19 & 20 UHF switches intercom, set and power selector switches
21 Collimator switch
22 ILS and radio frequency change card holders
23 UHF controller
24 Gun scavenging indicators
25 Gyroscopic Gun Sight (GGS)
26 Trigger safety catch
27 Gun trigger

28 Cine camera button
29 GGS spare bulb holders
30 VHF volume control
31 Armed time indicator Mk 2
32 Cine camera master switch
33 Cine camera sunny/cloudy switch
34 Stand-by UHF GD/A switch

This diagram from the manual shows the rear cockpit of the F(AW) Mk 9.
© Crown Copyright

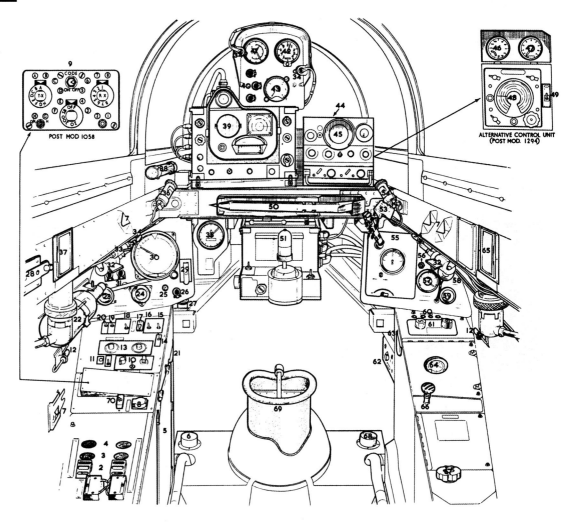

Key

Navigational Equipment

9 Rebecca Mk 8 control unit
46 Compass bearing indicator
6 FLS interrogation switch (foot operated)
5 Chart board
60 Pencil stowage
45 Gee indicator
19 Gee on/off switch
18 Gee and IFF standby switch
65 ADF card holder
30 Mk 4B compass master indicator
62 Compass control indicator lamps
40 AYF limit indicator lamps
43 AYF altitude indicator
41 Airspeed indicator
42 Altimeter
48 Radio compass receiver controller
49 Radio compass on/off switch
37 Rebecca card holder
70 Tail navigation lamp control switch

Radar

39 AI.17 indicator unit type 6744
51 AI.17 control unit type 901
61 AI.17 control unit, type 902
16 AI.17 inverters start switch
15 AI.17 inverters stop switch
14 AI.17 on/off switch
17 AI.17 and Gee DC circuit breakers
47 Range and heading meter
20 FIS on/off switch
26 FIS indicator lamp
11 IFF on/off switch and indicator lamp
10 IFF control unit
13 IFF coder controller
69 Visor stowage (for indicator visor [39])

Wireless

68 Press-to-mute switch (foot operated)
8 Press-to-transmit switch
56 Emergency intercom, push switch and lamp
25 Audio recorder off/stand-by switch

Hood [canopy] Controls

29 Hood control switch
33 Hood clutch release
52 Hood jettison handle
53 Hood jettison safety lock pin
63 Hood jettison pin stowage

Air Conditioning

32 Air conditioning louvres
67 Air ventilated suit on/off control

Oxygen

55 Oxygen regulator
66 Emergency oxygen release

Interior Lighting

12 Console lamps
59 Console lamps dimmer switch
36 Port chart board lamp
24 Port chart board lamp dimmer switch
54 Starboard chart board lamp
58 Starboard chart board lamp dimmer switch
34 Panel lamps
23 Panel lamps dimmer switch

Electrical Equipment

31 Generator warning lamps
2 Generator ground test switches
3 Generator voltmeter sockets
1 Generator voltage regulator remote trimmers
4 Generator ammeter sockets

Miscellaneous

50 Crow bar
38 Torch
22 Hand fire extinguisher
44 First-aid kit and asbestos gloves
27 Anti-G valves
57 Air temperature gauge
21 Stowage for seat additional firing handle safety pin
7 Harness clips
35 Fuel flowmeter
28 Seat firing pin stowage
64 Clock

F(AW) Mks 2, 6, and 8 all carried the American AI.22 radar, with the nose cone hinging to the left for maintenance, whereas AI.17 equipped aircraft required the whole nose cone to be completely removed.

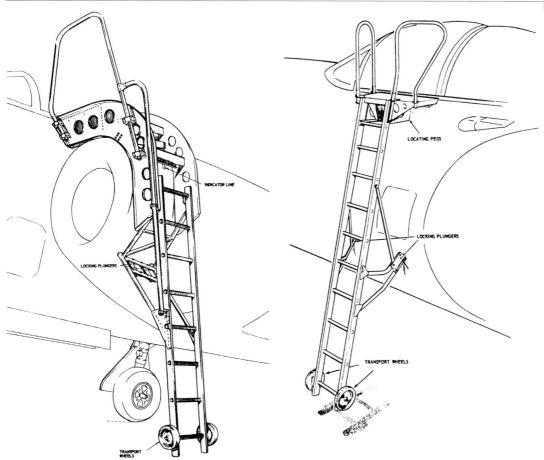

Far left: This is the rather bulky production crew access ladder used for all versions of the Javelin in RAF service.
© *Crown Copyright*

Left: With its extended nose, the T Mk 3 needed a revised access ladder for the front cockpit as shown here. The standard unit was used for the rear cockpit.
© *Crown Copyright*

The refuelling probe on a Javelin was a modification that could be added or taken away in the field. This diagram shows the parts necessary to fit the probe to the F(AW) Mk 9.
© Crown Copyright

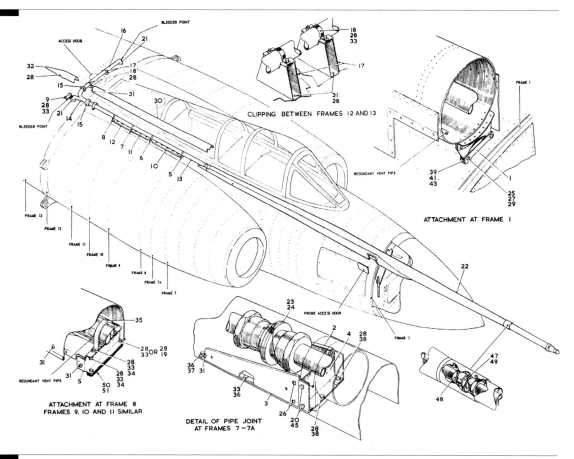

CLIPPING BETWEEN FRAMES 12 AND 13

ATTACHMENT AT FRAME 1

ATTACHMENT AT FRAME 8
FRAMES 9, 10 AND 11 SIMILAR

DETAIL OF PIPE JOINT
AT FRAMES 7 – 7A

This image of the wing from a T Mk 3 is appropriate for all early versions of the Javelin.
© Crown Copyright

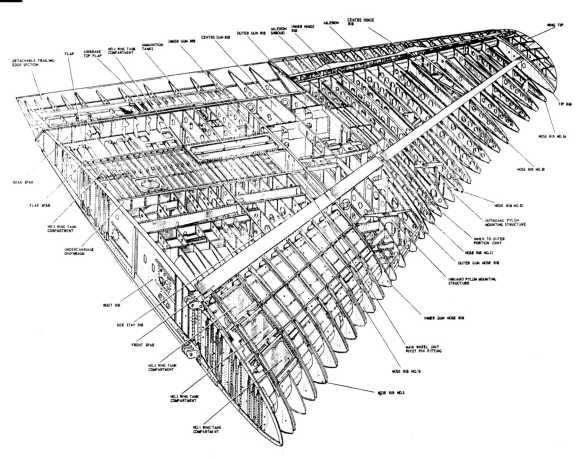

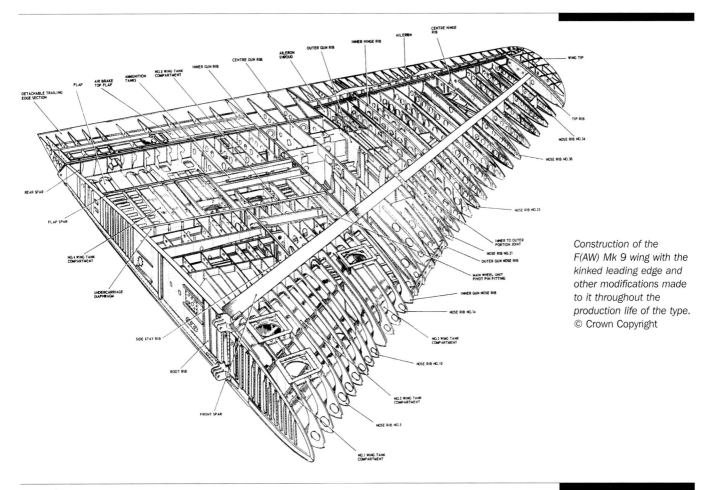

Construction of the F(AW) Mk 9 wing with the kinked leading edge and other modifications made to it throughout the production life of the type.
© Crown Copyright

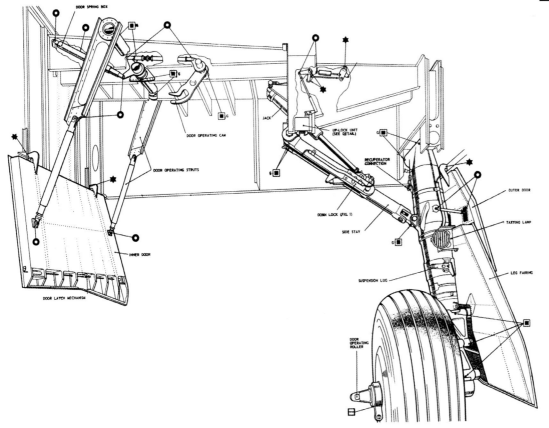

The undercarriage of the Javelin was sturdy, but not over complex as this diagram shows.
© Crown Copyright

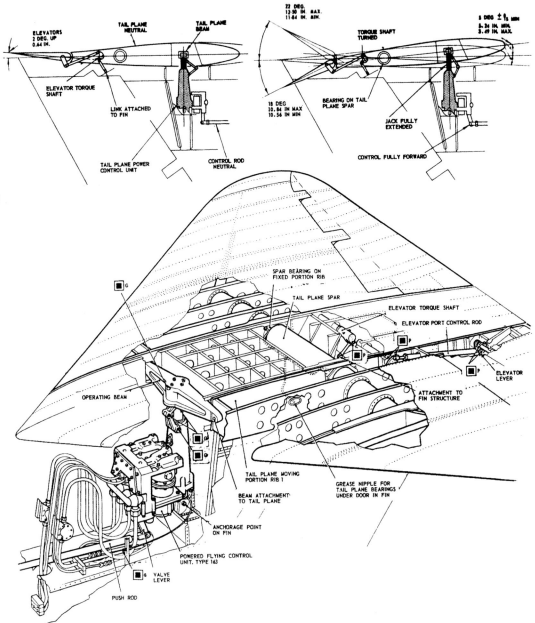

The actuator systems installed on the all-moving tail of the later series of the Javelin. You can also see the limits of the elevator movement.
© Crown Copyright

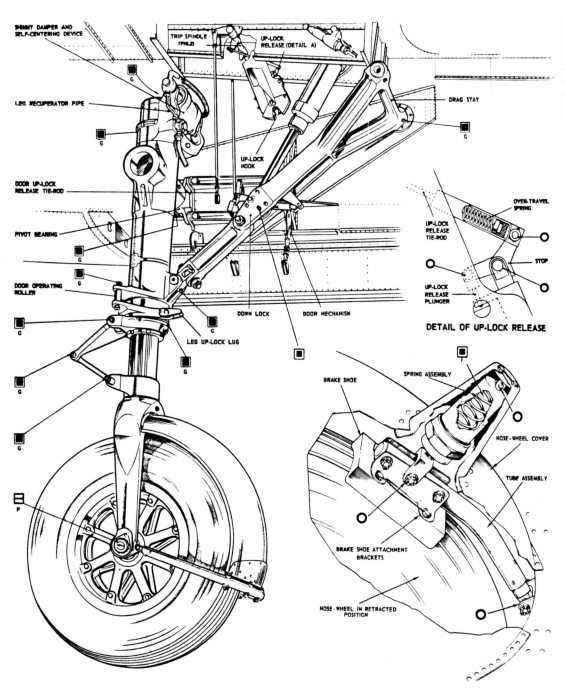

SHIMMY DAMPER AND
SELF-CENTERING DEVICE

TRIP SPINDLE
(FIG.2)

UP-LOCK
RELEASE (DETAIL A)

LEG RECUPERATOR PIPE

DRAG STAY

UP-LOCK
HOOK

DOOR UP-LOCK
RELEASE TIE-ROD

PIVOT BEARING

DOOR OPERATING
ROLLER

DOWN LOCK

DOOR MECHANISM

LEG UP-LOCK LUG

OVER-TRAVEL
SPRING

UP-LOCK
RELEASE
TIE-ROD

STOP

UP-LOCK
RELEASE
PLUNGER

DETAIL OF UP-LOCK RELEASE

BRAKE SHOE

SPRING ASSEMBLY

NOSE-WHEEL COVER

TUBE ASSEMBLY

BRAKE SHOE ATTACHMENT
BRACKETS

NOSE-WHEEL IN RETRACTED
POSITION

The nose leg fitted to all versions of the Javelin. The inset shows the buffer block that stopped the wheel spinning once the oleo had retracted into the nose well.
© Crown Copyright

Taken from the second Javelin production batch F(AW) Mk 4, XA632 was used for trials with four Firestreak missiles which were to become standard on F(AW) Mk 7, 8 and 9s. For these trials, an additional nose-mounted pitot head boom was installed.

The first production Javelin F(AW) Mk 4, XA629, was retained by Gloster Aircraft for various trials, among them being the fitting of Kuchemann streamlined aerodynamic bodies on the wing trailing edge, to reduce airflow separation at high subsonic speed and extend the buffet boundary. Two rows of vortex generators on the wing upper surface became standard on subsequent production aircraft.

Six earlier mark Javelins. An unusual photograph in its own right, but included here for the modeller perhaps. At first glance all of the aircraft camouflage schemes are identical, but closer scrutiny will identify slight variations.
Newark Air Museum

XA628, the final production F(AW) Mk 1. This aircraft was delivered to No.46 Squadron at RAF Odiham.

Chapter 6: PROJECTS, PROPOSALS AND DRAWING BOARD JAVELINS

Many projects were considered both during the initial design stages of the Javelin and for the later development and evolution of the type. What follows is a list of all those associated with the Javelin throughout both periods.

January 1947 - Specifications F.43/46 and F.44/46 issued for a single-seat day and two-seat night fighter.

P.228. A March 1947 design study for a two-seat night-fighter that bore a striking resemblance to the Meteor. It was armed with four 30mm ADEN cannon in the lower fuselage.

P.234. A February 1947 proposal that was an evolution of the Meteor with a delta wing and delta V-tail for a single-seat day fighter in response to Specification F.43/46. It was powered by two Rolls-Royce AJ.65 engines buried in the fuselage. A 4.5in recoilless gun was mounted under the fuselage.

P? (Delta Wing). This 1947 proposal does not have a 'P' number that is known, but it was similar to the P.234 only much larger and reverted to a standard tail unit.

P.238. A March 1947 proposal in relation to Specification F.43/46 and F.44/46 that had a delta wing, slab tail, was powered by two Metrovick F.9 Sapphires and armed with four 30mm cannon.

P.240. A proposal dated 18th April 1947 for a single-seat day and night interceptor to Specification F.43/46. This was a 30° swept wing version of P.238.

P.248. This proposal from August 1947 had lost all trace of the Meteor and now was a 'plank-winged' delta day and night interceptor to meet Specification F.43/46. This version was armed with a single 4.5in recoilless gun mounted in the nose.

P.250. This was the other version of the August 1947 proposal and was similar to the P.248, but was armed with two 30mm ADEN cannon in the nose.

P.258. This was a revision of the P.248 and P.250 projects brought about by the revision of F.43/46 and the call for the installation of radar and Red Hawk air-to-air missiles. The P.258 carried a Red Hawk in a semi-recessed bay below the fuselage. Another version had six rocket projectiles in the wing roots in lieu of the Red Hawk.

P.259. A version of the P.258 but with the 4.5in recoilless gun fitted in the port wing.

P.272. This was Gloster's first official proposal to F.4/48 and was dated 15th April 1948. The type looked very much like a Javelin and one version had four 30mm ADEN cannon, while another had four 4.5in recoilless guns.

P.275. A proposal from April 1948 for a single-seat day fighter to Specification F.3/48 which looked very similar to the Martin-Baker F.43/46 proposal. This was a delta 'dart' that also bore a strong resemblance to the Lippisch P.13a proposal from WWII.

P.276. A 1948 design tender to Specification F.4/48 for a two-seat night fighter. This design is believed to be the first that had the majority of features that would become the GA.5.

P.279. This was a design study dated 5th July 1949 for a two-seat night fighter to Specification F.4/48. This follows on from P.272 and was a larger airframe to better meet the requirements of the specification. Two, three and four engine layouts were considered, before Gloster's settled on two in their final brochure presentation.

P.280. This design is believed to be the first that fully shows what would become the GA.5. The final brochure produced by Gloster in July 1948 to meet Specification F.4/48 for a two-seat day & night fighter was entitled the P.280. This type would have had the Sa2 Sapphires replacing the AJ.65 Avons and saw the deletion of the rotating tips in favour of normal control surfaces.

P.315. A 1950 design study for a two-seat long-range fighter very similar to the Javelin and built to F.4/48. It had a 52ft span, was 57ft long and had a projected weight of 39,500lb.

P.316. A 1950 design study for a single-seat long-range fighter, also to F.4/48 but without radar. It was similar to the P.315 but was 58ft 8in long and would weigh 38,700lb.

P.317. A 1950 design proposal for a fighter-bomber version of the Javelin armed with 4x 1,000lb bombs.

P.318. A 1950 scheme for a two-seat long-range rocket-armed fighter.

P.319. 1950 proposal for a dual-control Javelin, developed and produced as the T Mk 3.

P.322. A 1950 design study for a Javelin to be powered by two Sa50 Sapphires. It was to have been capable of Mach 0.94 at 50,000ft.

P.323. A July 1950 design study for a long-range version of the P.322 project.

P.324. A 1950 design study for a fighter-bomber and rocket-armed version of the P.322 project.

P.325. A March 1951 proposal for a high-altitude escort fighter built to F.4/48 and designed to escort the Vickers B.9/48 (Valiant).

P.346. A 1952 design study for a Photo-Reconnaissance version of the Javelin (Scheme 1).

P.347. A 1952 design study for a Photo-Reconnaissance version of the Javelin (Scheme 2).

P.348. A 1952 design study of the P.347 fitted with wing-mounted cameras.

December 1952 - Operational Requirement 234 issued January 1953 - Specification PR.118D and P issued.

P.350. Tender placed for a PR version of the Javelin to meet Specification PR.118D and P.

P.356. A July 1953 tender for a thin-wing development of the Javelin. The design was an aside to the PR.118 D and P which had called for a thin-wing PR version of the Javelin, as the Air Staff had identified that a fighter version may also be required as well. The type would be armed with four Red Dean, four Blue Jay or two each of these, plus four 30mm ADEN cannon and have an AI.18 radar system. Enlarged this became the P.370, 371 and 372 series in September 1954.

P.359. Design relating to the Javelin F(AW) Mk 1 and 2.

P.364. A September 1953 development of the thin-wing Javelin powered by Olympus engines.

P.368. 1954 proposal for the installation of four Blue Jay missiles on a Javelin.

Specification F.153D issued

P.370 (G.50/G.A.6). This was a September 1954 proposal based on the P.356 only a lot larger. All of the P.370, 371 and 372 were based on the same requirement for a thin-wing development of the Javelin powered by the Olympus in response to F.153D. This version was armed with four 30mm ADEN cannon. Construction actually started on this project.

P.371(G.50/G.A.6). This was a revised version of the P.370 but this was to have been armed with two Red Dean missiles.

P.372 (G.50/G.A.6. This was a revised version of the P.370 but this was to have been armed with four Blue Jay missiles.

P.376. A May 1956 development proposal for a supersonic development of the P.370/1/2 series to meet F.153D. This was the ultimate thin-wing Javelin and a really big fighter. Armed with two Red Dean missiles the type was projected to achieve Mach 1.82 at 36,000ft. This project was cancelled on the 31st May 1956.

P.382. 1957 design study for the Javelin F(AW) Mk 7 fitted with two Scorpion rocket motors in a ventral pack.

P.505. A 1960 project for a STOL tactical strike reconnaissance aircraft to GOR.2 that was based on the Javelin.

Note:
The 'Thin Wing' Javelin

This programme, also known as the 'Super Javelin', was a proposed development of the Javelin that would see the basic fuselage added to a new set of wings and engines. In its initial state it was to have been subsonic, so Specifications F.118D and P were drafted around it. The type was to be considered an entirely new aircraft, not just a developed Javelin and the first project proposing this idea was the P.356 in July 1953.

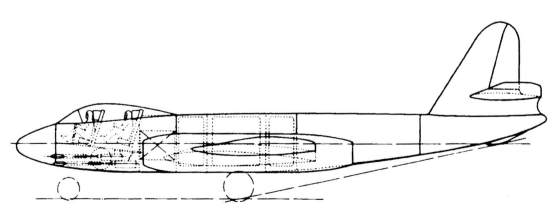

Produced in March 1948, the P.228 looked very much like a Meteor development.

Project 234 was created in February 1947 and had the delta wing but also incorporated a V-tail and the 4.5in recoilless gun.

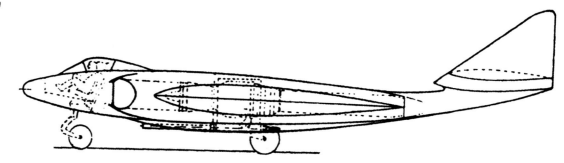

The first official submission in relation to Specification F.4/48. The design now looks much more like that which would become the Javelin and in this drawing the type features the 4.5in recoilless gun in the port wing and standard 30mm ADEN cannon in the starboard. This weapon layout was not incorporated in the final design.

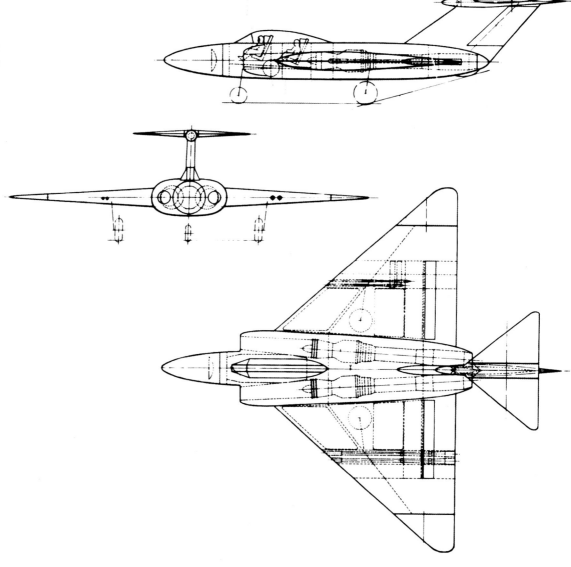

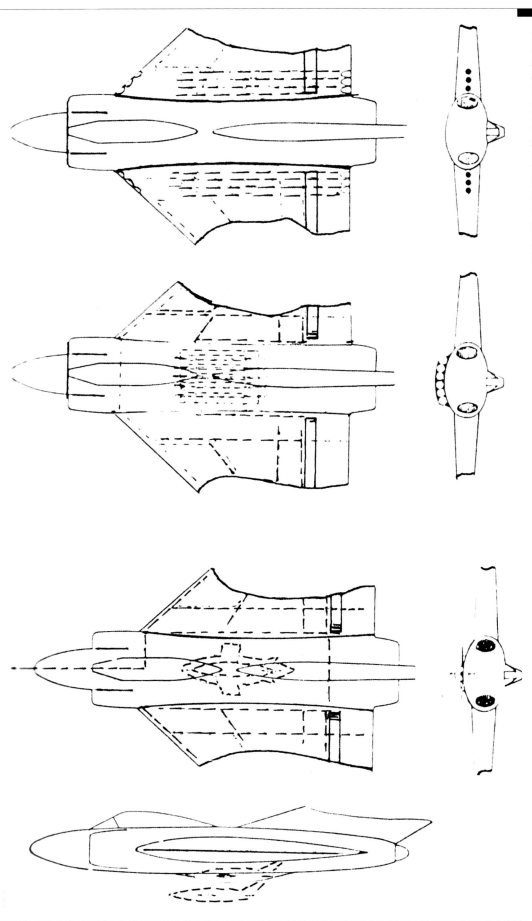

These drawings from October 1947 show the heavy armament initially considered for the Javelin. The upper one shows six rocket-projectiles fitted internally at the wing root, while the middle drawings show them fitted externally under the fuselage. The third drawing shows a rather interesting guided missile of a type not known, but similar in its almost aircraft-like form to those being considered at the end of WWII in Germany.

The German ancestry of this rather unorthodox design proposal from April 1948 is obvious, as the type looks very similar to the P.13a proposed in WWII by Dr Alexander Lippisch. The type was in response to Specification F.3/48.

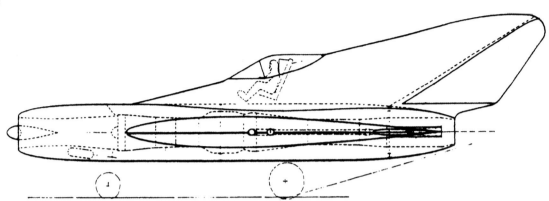

This is the 1950 proposal for a fighter-bomber version of the Javelin that could carry four 1,000lb freefall bombs.

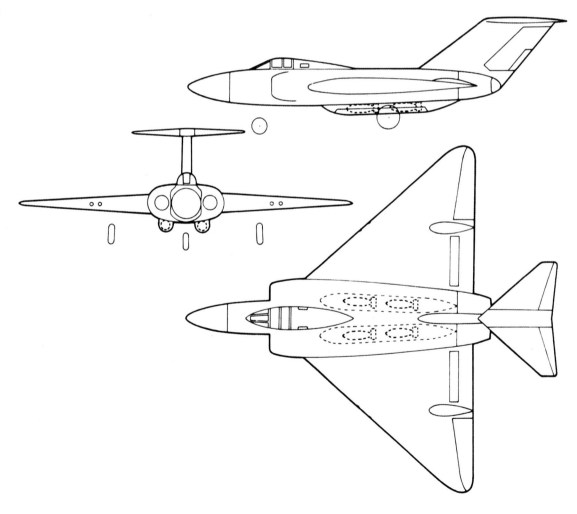

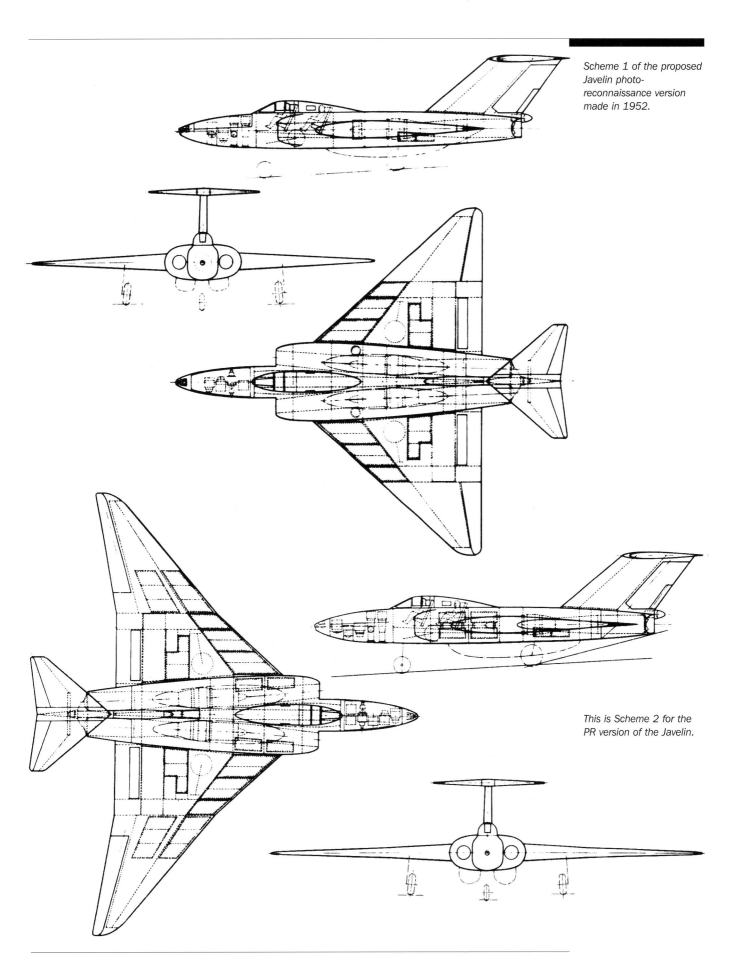

Scheme 1 of the proposed Javelin photo-reconnaissance version made in 1952.

This is Scheme 2 for the PR version of the Javelin.

Another 1952 PR proposal, this time with four cameras in the wings.

The PR Javelin proposal issued in relation to Specification PR.118D and P released in January 1953.

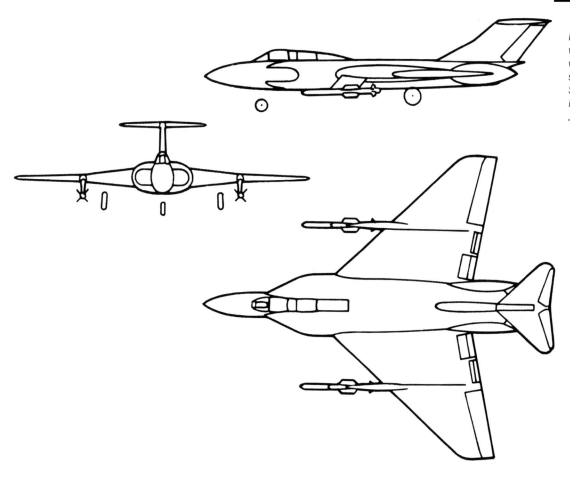

Proposal from May 1953 was for a supersonic version of the P.370/12 series designed to meet Specification F.153D. This is the ultimate 'thin-wing' Javelin development.

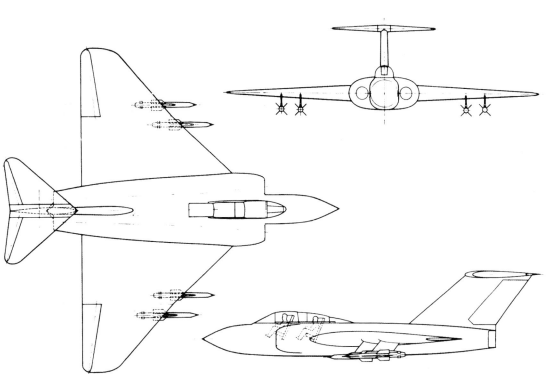

Drawing from April 1954 shows the Javelin Mk 1 and 2 fitted with four De Havilland Blue Jay missiles. These later went into production as the Firestreak.

The last Javelin incarnation, this proposal from 1960 was for a STOL tactical strike reconnaissance aircraft to meet specification GOR.2.

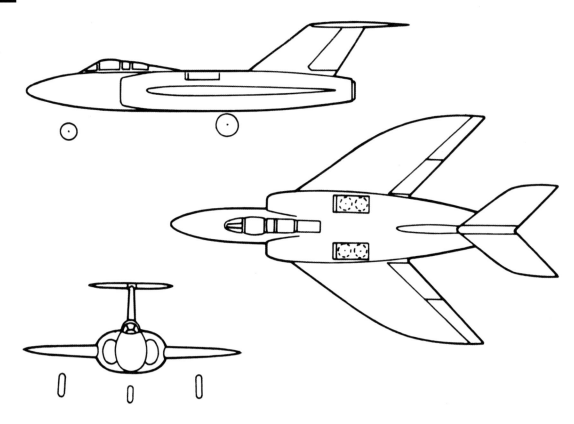

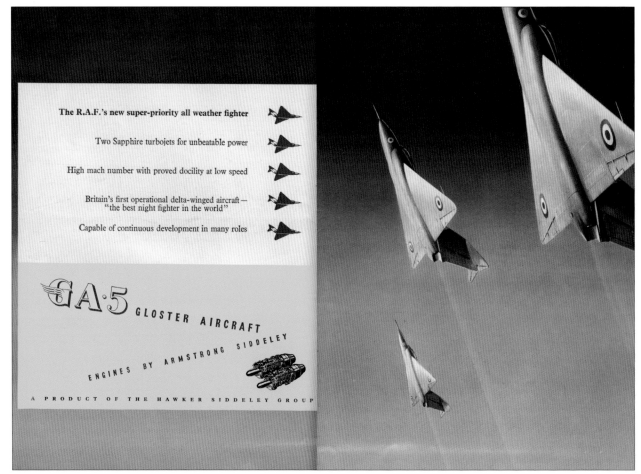

The R.A.F.'s new super-priority all weather fighter

Two Sapphire turbojets for unbeatable power

High mach number with proved docility at low speed

Britain's first operational delta-winged aircraft —
"the best night fighter in the world"

Capable of continuous development in many roles

GA·5 GLOSTER AIRCRAFT

ENGINES BY ARMSTRONG SIDDELEY

A PRODUCT OF THE HAWKER SIDDELEY GROUP

FACTS ABOUT THE GLOSTER JAVELIN / *Super Priority for the R.A.F.*

Take-off

TAKE-OFF at night or under conditions of limited visibility with a fully loaded aircraft can be a hazardous operation. The Javelin, possessed of a low wing loading and good power-weight ratio, is able to operate from exceptionally short runways and to climb steeply and swiftly to its combat altitude.

Gloster Aircraft Co. Ltd.
Hucclecote, Glos.

MEMBER OF THE HAWKER SIDDELEY GROUP / PIONEER . . . AND WORLD LEADER IN AVIATION

Chapter 7: **CAMOUFLAGE AND MARKINGS**

The Javelin, unlike many frontline aircraft, did not undergo any real changes in its camouflage and markings over the years, so what follows is a brief description of the schemes applied to the type in RAF service, both in the UK and abroad.

The Javelin's main camouflage scheme during its service life was a painted aluminium finish known as 'Speed Silver' on the undersurfaces and Dark Sea Grey with a disruptive camouflage pattern of Dark Green on the upper surfaces. The demarcation between upper and lower colours was quite low down on the fuselage sides. Knowing the RAF system I suspect that the point of demarcation was determined by identifying where a 60° angle would strike the fuselage sides. The Hycar radome was coated in Neoprene, so initially this was matt black, but in time it faded to a dark brown colour. This covering also seems to have been used on the intake lips of certain airframes, while others seem to be bare metal.

The roundels conformed with standard post-war national markings (introduced in late 1947) and widely referred to these days as the type 'D'. The roundels on either side of the nose below the cockpit were of 36in diameter, with a 48in version on the wing lower surface and 72in on the upper surface. The serial number was applied in 8in characters on the intake trunks below the front cockpit. Initially these were black, but at a later stage in the Javelin's career they were usually applied in white. The serial was also repeated in 36in black characters under the wings. These were orientated so that they could be read from the front under the port wing and from the rear for the

starboard. Usually an individual aircraft letter was applied in white characters high up on the vertical fin, although the positioning of this letter could vary considerably, often being dictated by the size or position of the squadron badges/fighter bands, assuming a letter was applied to the vertical tail at all. On some squadrons the individual letter was repeated on the forward fuselage close to the roundel. In some instances however the individual letter appeared only on the forward fuselage. The fin flash was formed by applying equal bands of red (front), white (centre) and blue (rear) to either side of the fin from the leading edge inwards and measured 72in high by 60in wide. The ventral 'bosom' fuel tanks were always finished in aluminium and the Firestreak missiles were usually painted one of two colours. A dummy round was black, while 'training' and live rounds were white.

There was one exception to RAF Javelins, this being F(AW) Mk 9 operated by No.228 OCU. The aircraft, XH898, was flown by Sqn Ldr George H Beaton and was bare metal overall. The radome remained black, the serials were black and all roundels and the fin flashes remained as those seen on camouflaged machines. The letters 'GHB' in 16in black characters were also on the fin, just aft of the fin flash. The intake lips were black and the ventral 'bosom' tanks were aluminium.

The prototypes all carried slight variations in the camouflage scheme in their lives. The first, WD804, initially appeared with a scheme of Dark Green and Dark Sea Grey on the upper surfaces and Medium Sea Grey underneath. However the camouflage demarca-

Visiting Little Rissington this view of F(AW) Mk 9, XH898, clearly shows her starboard side.
Ray Deacon

When flown by Sqn Ldr George H Beaton with No.228 OCU at Leuchars, F(AW) Mk 9, XH898 was in natural metal finish and carried his initials on the fin.

tion was situated halfway up the fuselage sides. The camouflage extended forward to cover the entire radome and the entire vertical fin was Medium Sea Grey. WD808 was in the same scheme, WT827 was the first to display the black radome, although it retained the same camouflage and demarcation as the previous prototypes. Some early aircraft however were used for test purposes and often had metal nose cones (not radomes) which were usually painted to match the aircraft's camouflage. The first aircraft to feature Speed Silver painted aluminium undersides was the first production Javelin F(AW) Mk 1 XA544.

F(AW) Mk 1 XA552 was used to test the DH Gyron Junior engines and was initially in standard camouflage colours, but was later painted Royal Blue overall (including the radome) with the legend 'The De Havilland Gyron Junior Javelin' in gold text on either side of the forward fuselage. The roundels on the nose were moved to the intake trunks and along with those on the wing and the fin flashes were outlined in white. The

serial was applied in white high up on the vertical fin, aft of the fin flash and in the usual RAF size and location under each wing, also in white. This aircraft usually carried 'bosom' tanks, and were also painted in Royal Blue.

The most colourful Javelin was F(AW) Mk 9 operated by the A&AEE. This aircraft XH897 was painted white overall with areas of day-glo red, that was applied in a complex pattern. Initially the type had a large red 'A' in a winged motif that had the words 'Fighter Test Squadron' running along the lower edge of the motif on either side of the nose in red. Later this was removed and a small 'A' was applied in red on either side of the nose within the white sections. See illustration on page 92-93.

The RAE also operated a very colourful example, F(AW) Mk 7, XH754. This aircraft was white overall with the nose and extreme rear fuselage in day-glo red. The entire undersurface, less the day-glo panels, was painted as per a target towing aircraft with equal

Having been employed by Gloster Aircraft as a revised-wing planform and ventral fuel tanks trials aircraft, the ninth production F(AW) Mk 1, XA552, ended its days painted an overall dark blue when used as a flying test-bed for two DH Gyron Junior DGJ.10 engines. These were destined for the Bristol Type 188, designed to explore kinetic heating effects on aircraft structures at Mach 2, over a sustained flight duration. As it was, the Gyron Junior's fuel consumption was far too excessive for the Type 188 to fulfil its designed role; the aircraft never exceeded Mach 1.8.

stripes of yellow and black running diagonally. This also applied to the underneath of the tailplane, although here it was black in the inner 2/3rd and yellow only on the outer 1/3rd with a straight demarcation line. On the tail the black did not extend to the elevators, which were yellow and the leading edge was also in yellow, even over the areas that were otherwise black. Where the serial numbers were crossed by the diagonal black stripes, the black was omitted, so that the serials looked as if they were in yellow 'blocks' The leading edges of the wings, vertical fin and tailplanes were in a buff colour.

Below: *Converted from F(AW) Mk 7 to F(AW) Mk 9 standard at Moreton Valence, XH897 is here displayed in September 1968, in a colourful paint scheme while on the strength of 'A' Squadron, the A&AEE's unit dedicated to the development of fighters and single-seat strike aircraft. In the squadron, XH897 was employed for pressure error measurement, with pressure heads on the nose cone and each wing tip.*

Bottom: *Devoid of any apparent serial number we believe this photograph originates from Glosters themselves. The aircraft would appear to be painted white overall with black anti-glare panels on the nose and engine nacelles.*

F(AW) Mk 1

Javelin F(AW) Mk 1, XA554, 'Exercise Beware', RAF Coltishall, whilst serving with the A&AEE at Boscombe Down in October 1955. The standard colour scheme for Javelins consisted of painted aluminium undersides known as Speed Silver, Dark Sea Grey with a disruptive camouflage pattern of Dark Green on the upper surfaces. Roundels in six positions with serials positioned on engine nacelles, repeated under the wings; note eight Canberra 'kills' marked under the windscreen. This is an early production Mk 1 with early style short nose and low profile pilot's canopy.

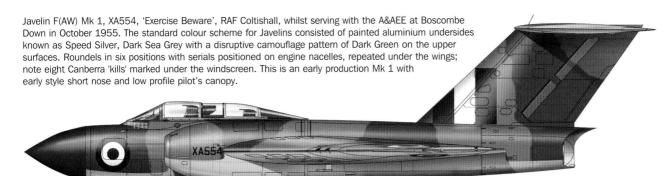

Javelin F(AW) Mk 1, XA568, College of Aeronautics, Cranfield. Standard scheme and national markings. Orange Day-glo air intakes and fin.

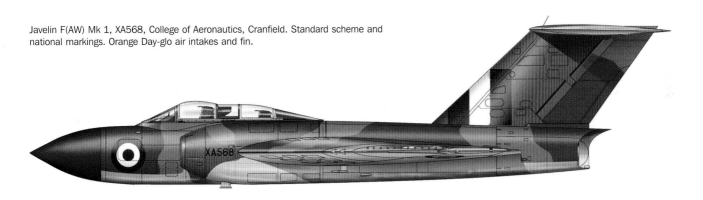

Javelin F(AW) Mk 1, XA623, 'G', No.46 Squadron (B Flight). Standard scheme and national markings. Unit flash in red and white under nose roundel, repeated on a white rectangle on fin. Yellow 'G' on fin.

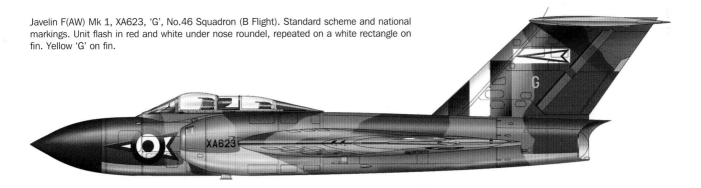

Javelin F(AW) Mk 1, XA618, 'N', No.46 Squadron (A Flight). Standard scheme and national markings . Unit flash in red on a white rectangle on fin. Red 'N' on fin.

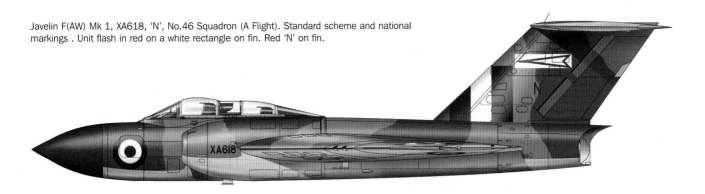

Javelin F(AW) Mk 1, XA558, 'A', No.87 Squadron, 2nd Tactical Air Force, RAF Brüggen, West Germany. Standard scheme and national markings, except for the fuselage roundel which has been reduced in size to fit within the unit's colour bar; note No.87's sword insignia on fin in red, outlined in white. Crashed near Brüggen on 5th June 1958 after both engines flamed out.

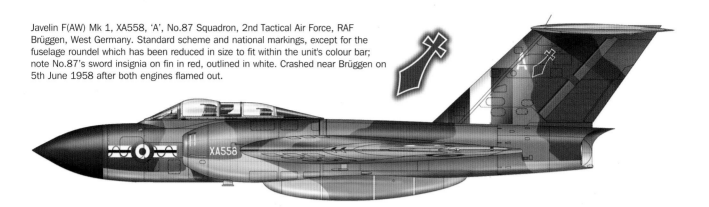

F(AW) Mk 2

Javelin F(AW) Mk 2, XA776, 'N', No.46 Squadron, RAF Odiham, 1957. Standard scheme and national markings. White and red arrow with an RAF roundel superimposed on fin. Yellow 'N' code.

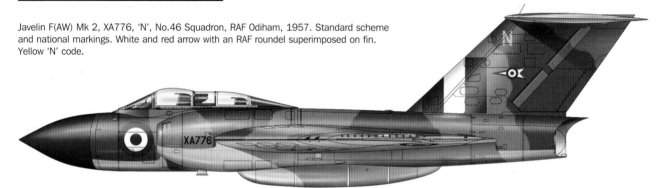

Javelin F(AW) Mk 2, XA774. 'J', No.89 Squadron, RAF Stradishall, 1958. Standard scheme and national markings. White code 'J' on fin above squadron colours of light and dark blue.

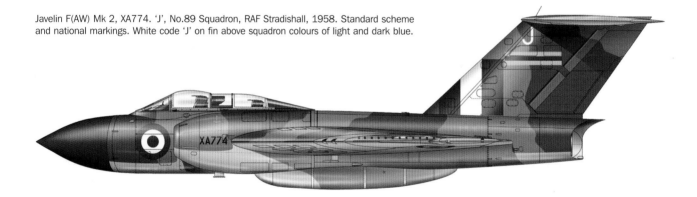

Javelin F(AW) Mk 2, XA780, 'C', No.46 Squadron, RAF Waterbeach, June 1961. Standard scheme and national markings. Yellow 'C' code on fin, with the unit badge in a white circle superimposed on a red/white arrowhead.

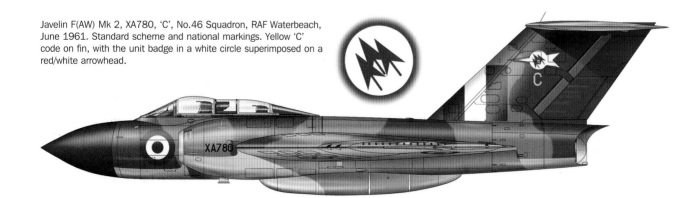

T Mk 3

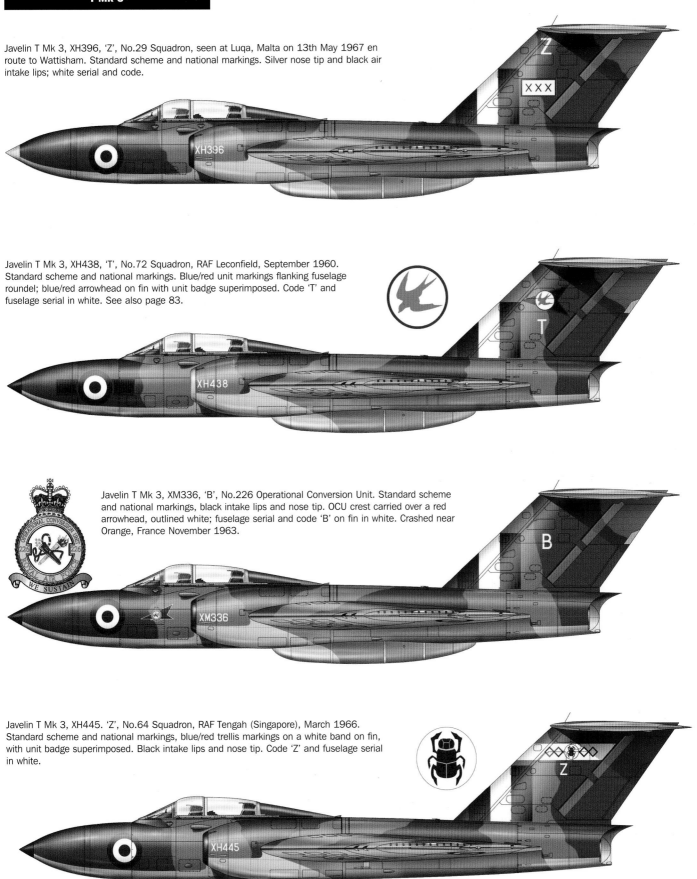

Javelin T Mk 3, XH396, 'Z', No.29 Squadron, seen at Luqa, Malta on 13th May 1967 en route to Wattisham. Standard scheme and national markings. Silver nose tip and black air intake lips; white serial and code.

Javelin T Mk 3, XH438, 'T', No.72 Squadron, RAF Leconfield, September 1960. Standard scheme and national markings. Blue/red unit markings flanking fuselage roundel; blue/red arrowhead on fin with unit badge superimposed. Code 'T' and fuselage serial in white. See also page 83.

Javelin T Mk 3, XM336, 'B', No.226 Operational Conversion Unit. Standard scheme and national markings, black intake lips and nose tip. OCU crest carried over a red arrowhead, outlined white; fuselage serial and code 'B' on fin in white. Crashed near Orange, France November 1963.

Javelin T Mk 3, XH445. 'Z', No.64 Squadron, RAF Tengah (Singapore), March 1966. Standard scheme and national markings, blue/red trellis markings on a white band on fin, with unit badge superimposed. Black intake lips and nose tip. Code 'Z' and fuselage serial in white.

Javelin T Mk 3, XH437, '3T3', No.33 Squadron, 1961. Standard scheme and national markings; white code '3T3' thinly outlined in yellow on both sides of nose, serial in black. Stripe along tail in light blue, dark blue and red. Silver intake lips and black nose tip. Written off at Leuchars in August 1964 following a fire.

Javelin T Mk 3, XH393, 'T3', No.5 Squadron, RAF Geilenkirchen, West Germany, summer 1965. Standard scheme and national markings. Black intake lips and nose tip, white serial and code; red/pale blue band across tail with unit badge in a white circle. Note reduced size of nose roundel.

Javelin T Mk 3, XH390, 'O', No.60 Squadron, RAF Tengah, Singapore, 1966. Standard scheme and national markings; black intake lips and nose tip. White serial and code; black/white stripes along tail.

Javelin T Mk 3, XH397, 'B', No.228 OCU (No.11 Reserve Squadron), 1966. Standard scheme and national markings. Black intake lips and nose tip. White code and serial, white disc on fin with No.11 Squadron marking in brown.

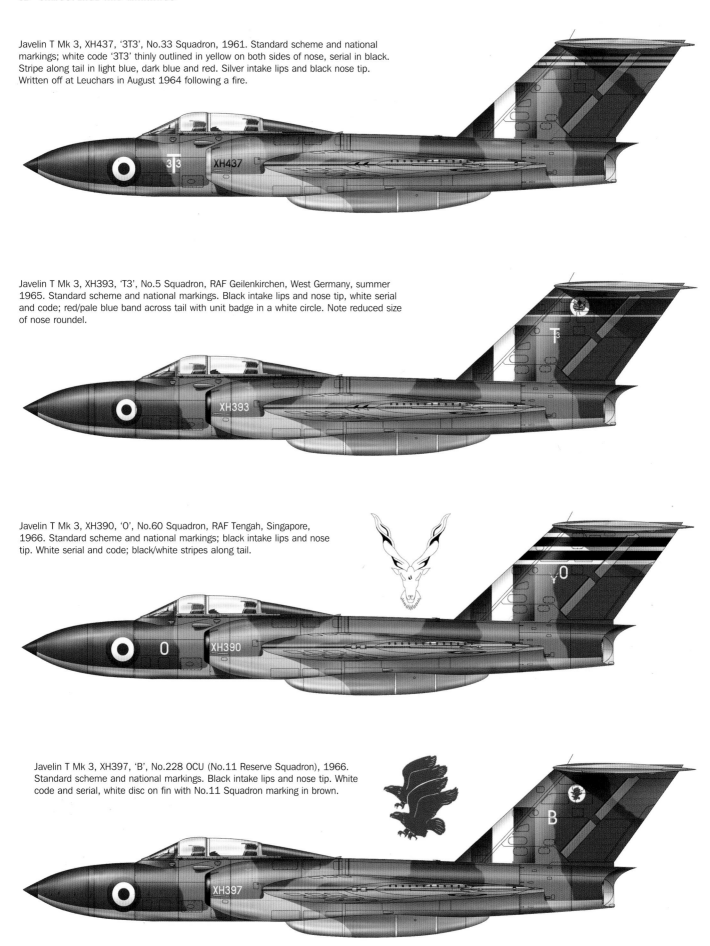

Javelin T Mk 3, XH436, 'R', No.151 Squadron, RAF Leuchars, September 1960. Standard scheme and national markings. Silver air intake lips and black nose tip. White serial and code; St Andrew's cross outlined in black. Written off at Leuchars in May 1964 following a fire.

Javelin T Mk 3, XH446, '06', No.228 OCU (No.137 Reserve Squadron). Standard scheme and national markings; white serial and code, silver air intake lips and black nose tip. Unit badge on fin.

Javelin T Mk 3, XH438, 'A', Fighter Command Instrument Rating Squadron. Standard scheme and national markings with white serial and code; FCIRS (in full) in white on nose, with Squadron Leader's pennant above it. Black air intake lips and black nose tip.

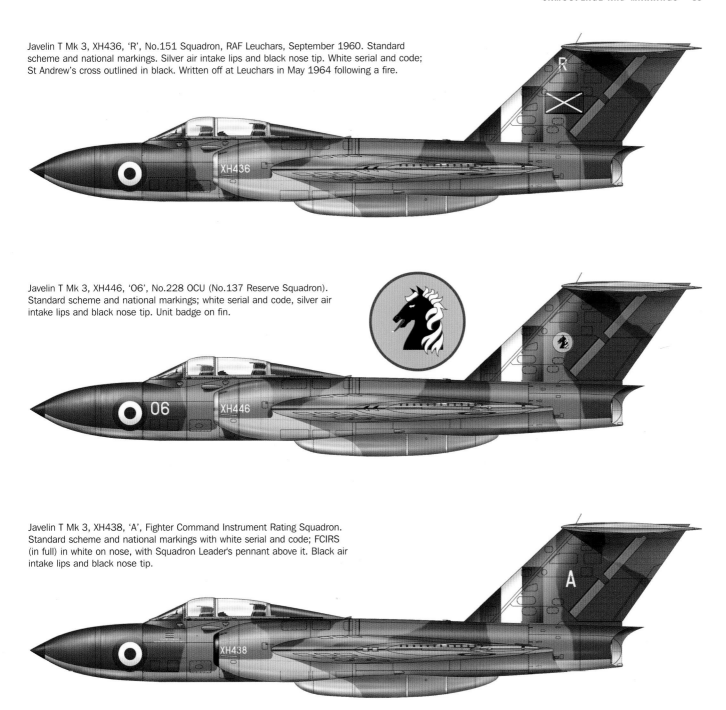

F(AW) Mk 4

Javelin F(AW) Mk 4, XA756, '756', No.141 Squadron, RAF Coltishall, 1957. Standard scheme and national markings; black/white fin markings with unit badge superimposed. Fuselage serial in black, with '756' repeated in white on fin.

Javelin F(AW) Mk 4, XA635, 'L', No.3 Squadron, RAF Geilenkirchen 1959. Standard scheme and national markings; Apple Green band around fin with unit crest superimposed. Fuselage serial and 'L' on fin in white.

Javelin F(AW) Mk 4, XA753, No.23 Squadron, 1957. Standard scheme and national markings; blue/red/blue bars above black serial on engine nacelles. Unit badge in red/black and yellow on fin.

Javelin F(AW) Mk 4, XA636, '636', No.41 Squadron, RAF Coltishall, 1958. Standard scheme and national markings; black serial on engine nacelles, with the last three digits repeated in white on fin. Note white bars only flanking unit badge on fin.

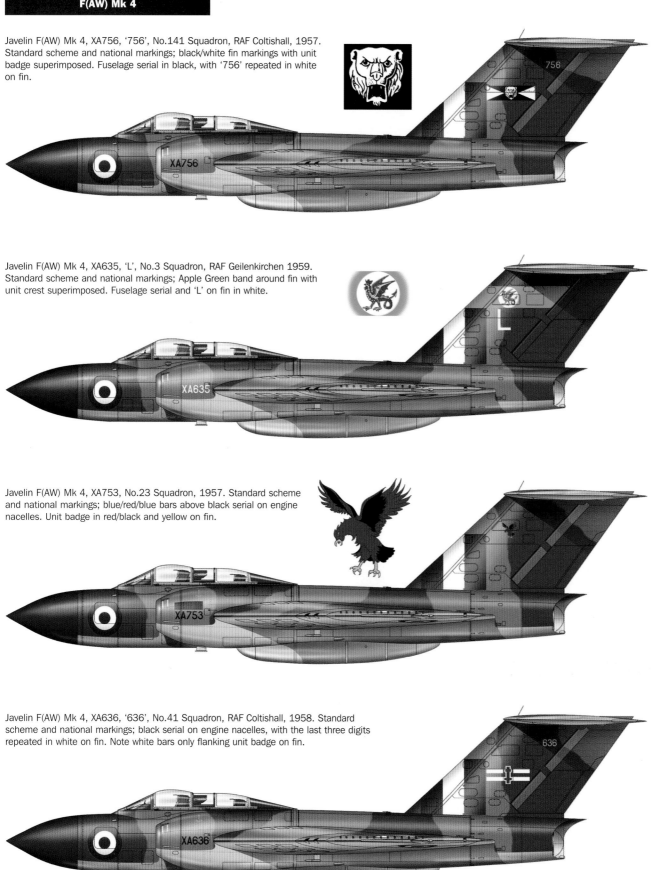

Javelin F(AW) Mk 4, XA632, 'A', No.11 Squadron, RAF Geilenkirchen 1960. Standard scheme and national markings; white serial. Yellow and black fin markings with black code on a white circle outlined in black.

Javelin F(AW) Mk 4, XA737, 'K', No.72 Squadron, RAF Leconfield, 1959. Standard scheme and national markings; white serial and code. Unit badge superimposed on a blue arrowhead outlined in red on fin.

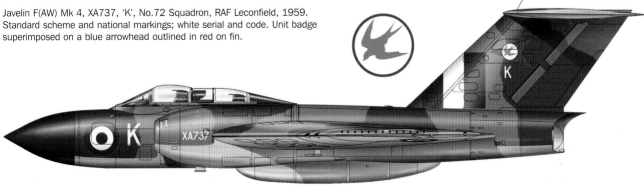

Javelin F(AW) Mk 4, XA730, 'N', No.72 Squadron, RAF Leconfield, 1960. Standard scheme and national markings; white serial and code. Blue squadron markings, edged in red, flank the fuselage roundel, colours repeated in arrow form on fin with the unit badge superimposed.

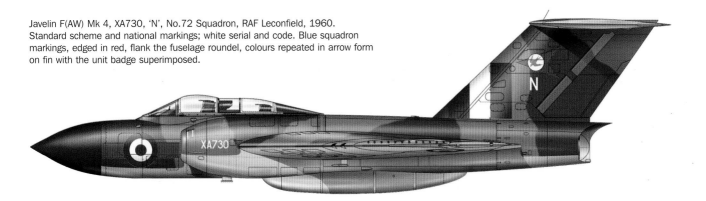

F(AW) Mk 5

Javelin F(AW) Mk 5, XA664, 'P', No.5 Squadron, RAF Laarbruch, West Germany, 1961. Standard scheme and national markings; red band across fin with unit badge within a white disc. Black intake lips; white serial and code. Ex-FCS aircraft.

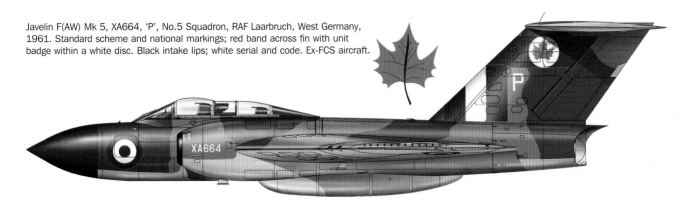

Javelin F(AW) Mk 5, XA654, No.23 Squadron, 1958 (the only Mk 5 on the squadron). Standard scheme and national markings; white serial with blue/red/blue bars above. Unit badge on a white disc on fin.

Javelin F(AW) Mk 5, XA707. 'A', No.41 Squadron. Standard scheme and national markings; white code and serial. Unit badge, flanked by white and red bars, on fin.

Javelin F(AW) Mk 5, XA710, 'Y', No.151 Squadron, RAF Leuchars. Standard scheme and national markings; St Andrew's cross outlined in black, on fin. Serial in black and code in white; note unit crest on fuselage.

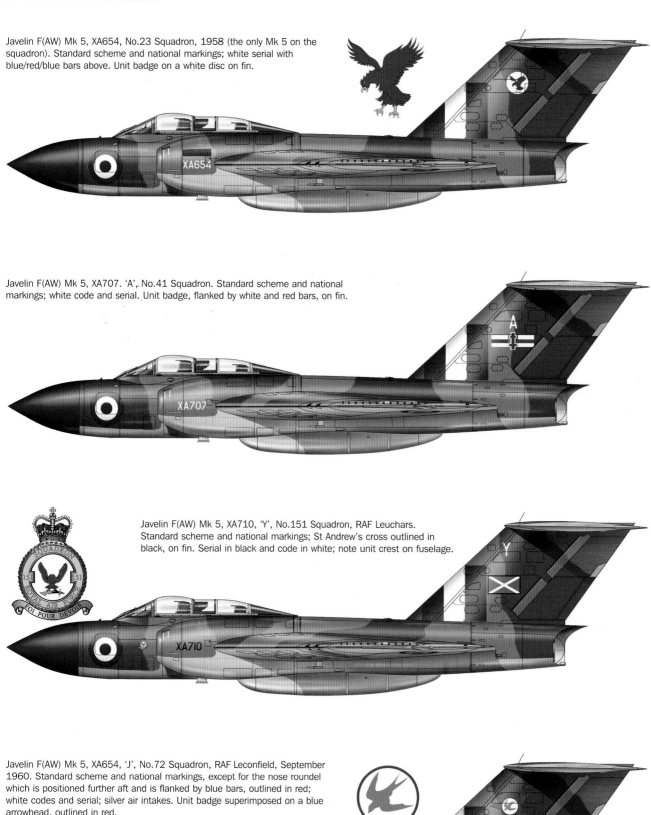

Javelin F(AW) Mk 5, XA654, 'J', No.72 Squadron, RAF Leconfield, September 1960. Standard scheme and national markings, except for the nose roundel which is positioned further aft and is flanked by blue bars, outlined in red; white codes and serial; silver air intakes. Unit badge superimposed on a blue arrowhead, outlined in red.

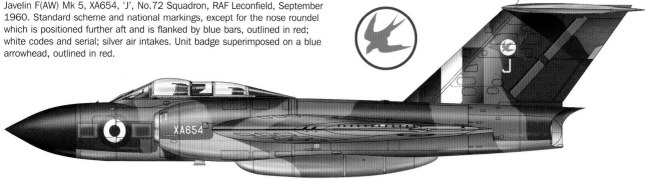

Javelin F(AW) Mk 5, XA652, 'T', All-Weather Fighter Combat School (No.219
Reserve Squadron), 1962. Standard scheme and national markings; white
serial and code. Orange day-glo wing tips.

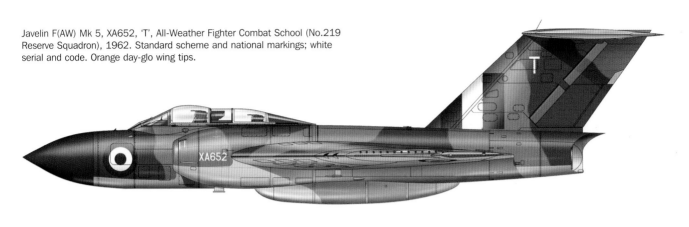

Javelin F(AW) Mk 5, XA667, 'O', No.228 OCU, early 1961.
Standard scheme and national markings; white serial and code.
Unit badge on fin.

F(AW) Mk 6

Javelin F(AW) Mk 6, XA817, 'E', No.29 Squadron. Standard scheme and national markings;
black serial and white code. white bar, outlined in red, with three 'X' in red, on fin.

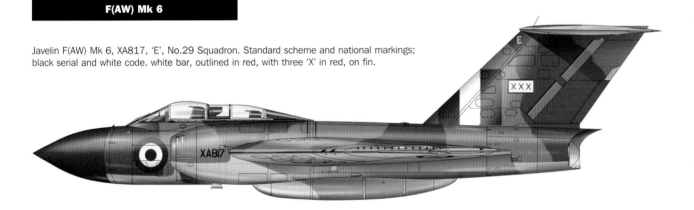

Javelin F(AW) Mk 6, XH694, 'A', No.85 Squadron, RAF West Malling, October 1959.
Standard scheme and national markings; black/red checks on fin with unit badge in white
superimposed. Serial in black; 'A' on fin in yellow.

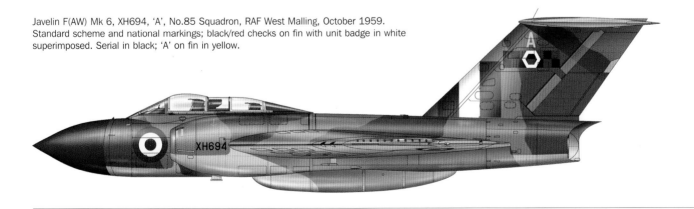

Javelin F(AW) Mk 6, XA815, 'E', No.89 Squadron, RAF Stradishall, 1958. Standard scheme and national markings; black serial and white code. Light/dark/light blue bars on fin; note extended gun barrels.

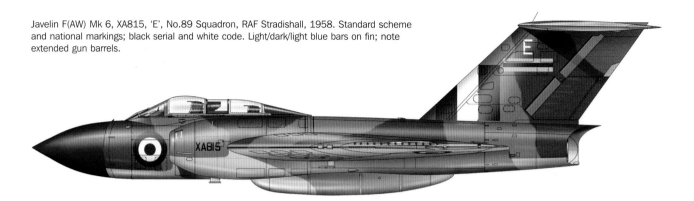

F(AW) Mk 7

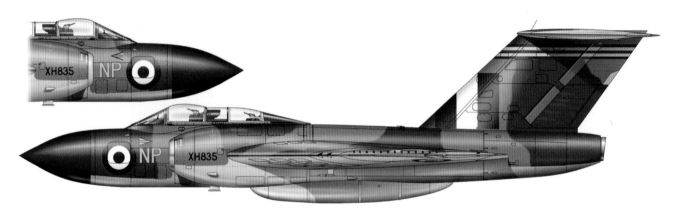

Above: Javelin F(AW) Mk 7, XH835, 'NP', No.33 Squadron. Standard Scheme and national markings; black serial and silver air intake lips. Code in light blue, outlined in red; light blue, red and blue tail stripes. Wing Commander's pennant above code on nose (note slightly different position on the starboard side).

Below: Javelin F(AW) Mk 7, XH794, 'L', No.64 Squadron, RAF Duxford. Standard scheme and national markings; white serial and code. Silver air intake lips; white band on fin with red and blue trellis and unit badge within a white disc, the latter outlined in blue.

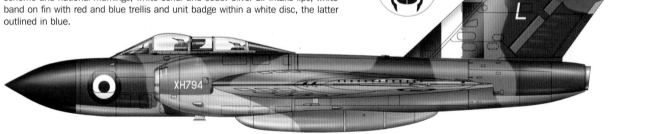

Javelin F(AW) Mk 7, XH899, 'P', No.25 Squadron as seen at Sculthorpe, 16th May 1959. Standard scheme and national markings; white serial and code, the latter repeated on fuselage. Silver and black bands flank squadron crest on fin.

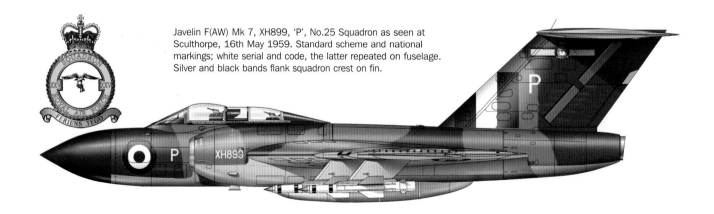

Javelin F(AW) Mk 7, XH778, 'G', No.23 Squadron. Standard scheme and national markings; white code and serial. Silver air intakes; blue/red/blue bars, outlined in white, above serial on engine nacelles. Unit badge on fin within a white disc.

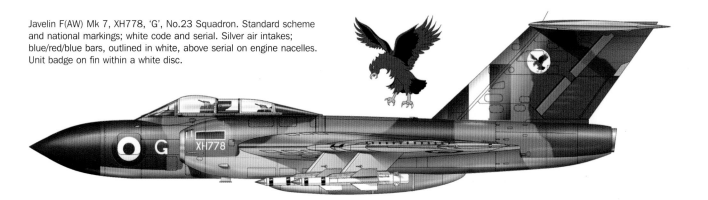

F(AW) Mk 8

Javelin F(AW) Mk 8, XH966, 'X', No.41 Squadron, RAF Wattisham, September 1963. Standard scheme and national markings; red/white bars flanking unit marking on fin. Serial and fin code in white; air intake lips are black.

Javelin F(AW) Mk 8, XJ123, 'B', No. 85 Squadron, 1962. Standard scheme and national markings; white serial and code with black air intake lips. White hexagon (non-standard, elongated) on black/red checks on fin. Note: hexagon extends slightly above and below the chequered band as opposed to XH694 shown on page 87.

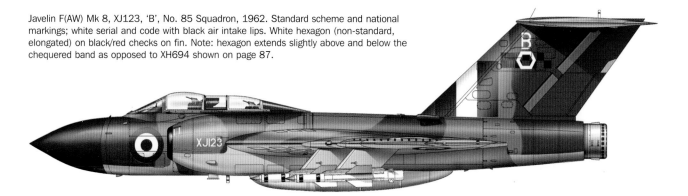

Javelin F(AW) Mk 8, XH979, Air Fighting Development Squadron, RAF Binbrook, 1962. Standard scheme and national markings; AFDS written in full in white under crest on fin. White fuselage serial; black air intake lips.

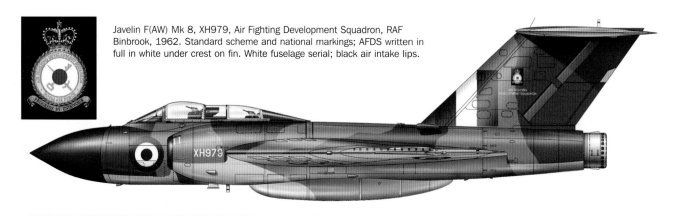

F(AW) Mk 9 & F(AW) Mk 9R

Javelin F(AW) Mk 9R, XH780, 'A', No.5 Squadron, RAF Geilenkirchen. Standard scheme and national markings; white serial and code. Black air intake lips; red/pale blue band on tail with unit badge superimposed. (Squadron markings now incorporating part of No.33 Squadron's colours following the disbandonment of No.33 Squadron and the transfer of its aircraft to No.5 Squadron in November 1962).

Javelin F(AW) Mk 9, XH772, 'G', No.11 Squadron, RAF Geilenkirchen (Germany). Standard scheme and national markings; white serial and code. Black air intake lips; yellow/black band across tail with unit marking superimposed.

Javelin F(AW) Mk 9R, XH793, 'A', No.23 Squadron, RAF Coltishall, late 1960. Standard scheme and national markings; serial and code in white. Unit marking in a white disc on fin; black air intake lips.

Javelin F(AW) Mk 9, XH882, 'L', No.25 Squadron. Standard scheme and national markings; white serial and code. Black intake lips; silver bars, edged in black, flanking unit crest on fin.

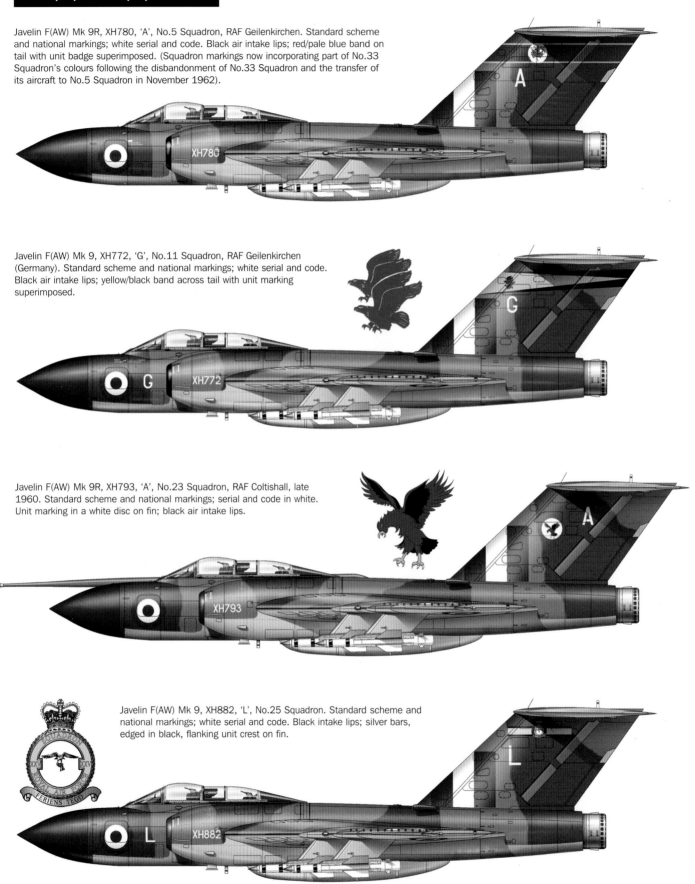

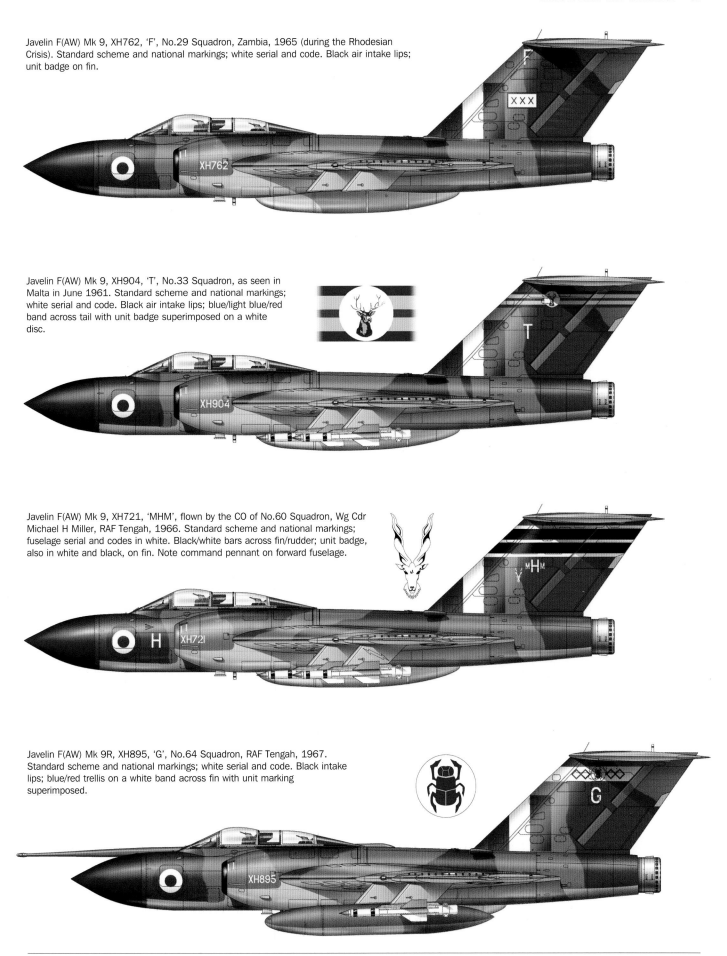

Javelin F(AW) Mk 9, XH762, 'F', No.29 Squadron, Zambia, 1965 (during the Rhodesian Crisis). Standard scheme and national markings; white serial and code. Black air intake lips; unit badge on fin.

Javelin F(AW) Mk 9, XH904, 'T', No.33 Squadron, as seen in Malta in June 1961. Standard scheme and national markings; white serial and code. Black air intake lips; blue/light blue/red band across tail with unit badge superimposed on a white disc.

Javelin F(AW) Mk 9, XH721, 'MHM', flown by the CO of No.60 Squadron, Wg Cdr Michael H Miller, RAF Tengah, 1966. Standard scheme and national markings; fuselage serial and codes in white. Black/white bars across fin/rudder; unit badge, also in white and black, on fin. Note command pennant on forward fuselage.

Javelin F(AW) Mk 9R, XH895, 'G', No.64 Squadron, RAF Tengah, 1967. Standard scheme and national markings; white serial and code. Black intake lips; blue/red trellis on a white band across fin with unit marking superimposed.

Javelin F(AW) Mk 9, XH716, 'G', No.228 OCU (No. 11 Reserve Squadron), 1966. Standard scheme and national markings; white serial and code. Black air intake lips; No.11 Squadron badge here rendered in black on a white disc. The birds having also appeared in brown on occasion! See pages 82 and 90.

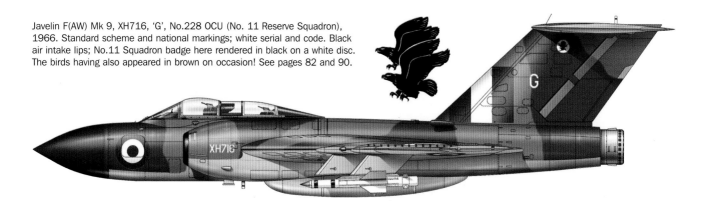

Javelin F(AW) Mk 9, XH898, 'GHB', flown by the CO of No.228 OCU, RAF Leuchars, Sqn Ldr George H Beaton. Known to be the only natural metal finished Javelin to see operational service; black cockpit framing, intake lips and serials; standard national markings. Pilot's initials on fin in black.

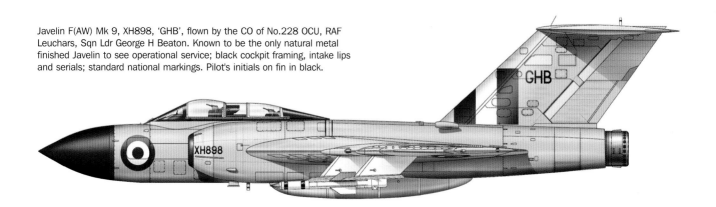

Below and opposite page: Javelin F(AW) Mk 9, XH897, 'A', of the A&AEE Boscombe Down, September 1970. Red/white scheme, with black anti-dazzle panel and cockpit framing. Standard national markings; serials and intake lips in black. 'A' on nose also in red over an arrowhead.

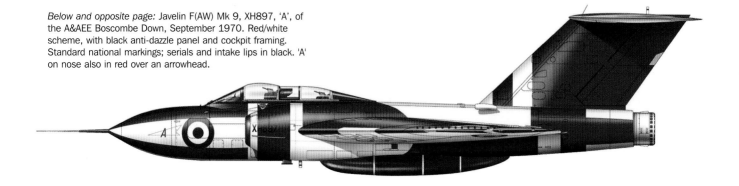

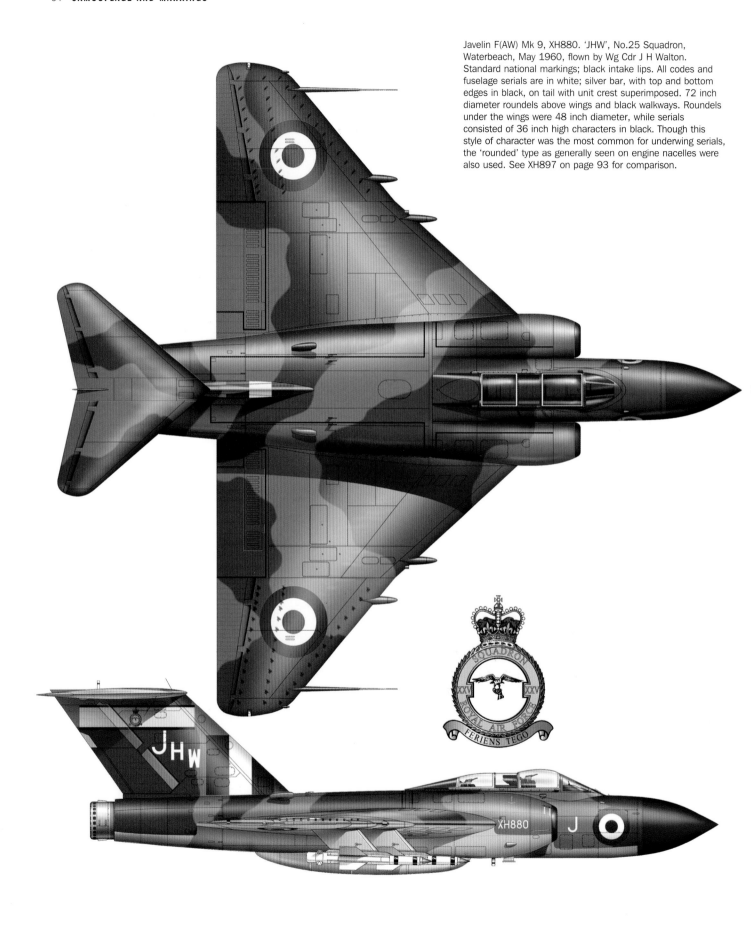

Javelin F(AW) Mk 9, XH880. 'JHW', No.25 Squadron, Waterbeach, May 1960, flown by Wg Cdr J H Walton. Standard national markings; black intake lips. All codes and fuselage serials are in white; silver bar, with top and bottom edges in black, on tail with unit crest superimposed. 72 inch diameter roundels above wings and black walkways. Roundels under the wings were 48 inch diameter, while serials consisted of 36 inch high characters in black. Though this style of character was the most common for underwing serials, the 'rounded' type as generally seen on engine nacelles were also used. See XH897 on page 93 for comparison.

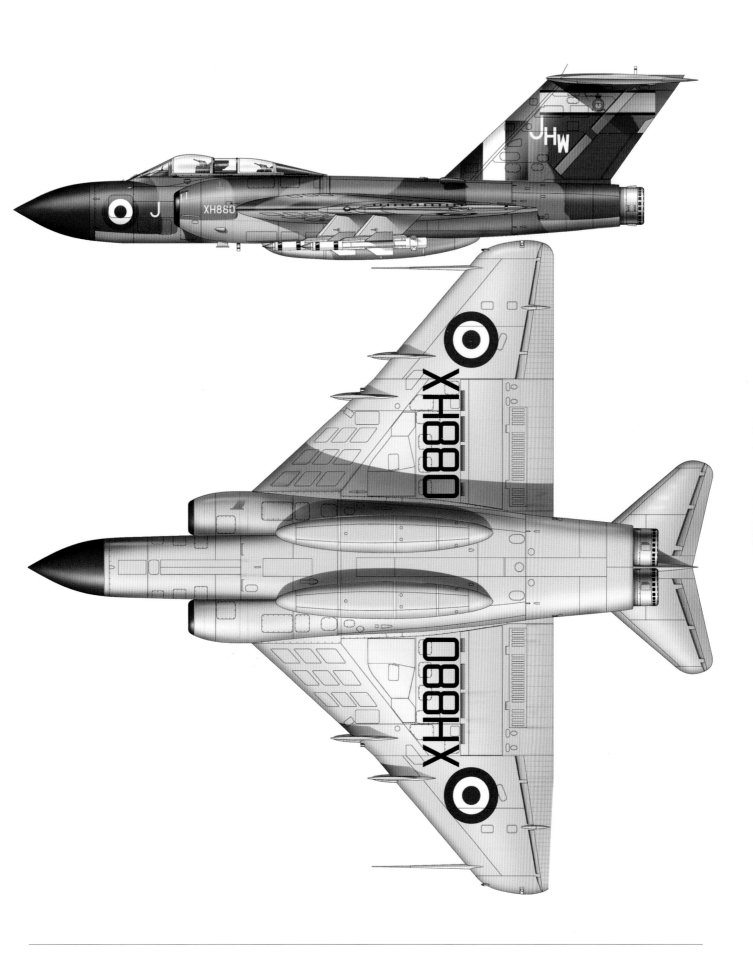

Javelin F(AW) Mk 2, XA778, A&AEE Boscombe Down, September 1967. Day-glo Orange overall with 'dirty' yellow rear fuselage and port ventral tank; starboard ventral tank is natural metal. Black anti-dazzle panel, air intake lips and cockpit framing. This aircraft was a hybrid, being basically a F(AW) Mk 2 with F(AW) Mk 7 rear fuselage and engine and F(AW) Mk 8 flying controls.

Appendix I: **TECHNICAL DATA**

Fighter (All-Weather)

Designation: F(AW) Mk 1

First Flight: 22nd July 1954 (XA544)
Span: 52ft 0in
Length: 56ft 3in
Height: 16ft 0in
Engine: Two Armstrong-Siddeley Sapphire Sa6 Mk 10201 (port) and 10301 (starboard) turbojets offering 8,300lb thrust.
Fuel Capacity: 765 imperial gallon plus capacity to carry two 250 imperial gallon ventral tanks.
Weight: Take-off (clean) 31,580lb, Overload 36,690lb (two ventral tanks).
Max speed @Sea Level (clean): 700mph, at 40,000ft 614mph.
Climb to 45,000ft: 9.8 minutes
Ceiling: Service 52,500ft, Absolute 55,000ft.
Armament: Four fixed 30mm ADEN cannon mounted in the outer wing panels.
Production: 40

Designation: F(AW) Mk 2

First Flight: 31st October 1955 (XD158)
Span: 52ft 0in
Length: 56ft 3in
Height: 16ft 0in
Engine: Two Armstrong-Siddeley Sapphire Sa6 Mk 10701 (port) and 10801 (starboard) turbojets offering 8,300lb thrust.
Fuel Capacity: 765 imperial gallon plus capacity to carry two 250 imperial gallon ventral tanks.
Weight: Take-off (clean) 32,100lb, Overload 37,200lb (two ventral tanks).
Max speed @Sea Level (clean): 700mph, at 40,000ft 614mph.
Climb to 45,000ft: 9.8 minutes
Ceiling: Service 52,500ft, Absolute 55,000ft
Armament: Four fixed 30mm ADEN cannon mounted in the outer wing panels.
Production: 30

Designation: F(AW) Mk 4

First Flight: 19th September 1955 (XA629)
Span: 52ft 0in
Length: 56ft 3in
Height: 16ft 0in
Engine: Armstrong-Siddeley Sapphire Sa6 Mk 10201 (port) and 10301 (starboard) turbojets offering 8,300lb thrust.
Fuel Capacity: 765 imperial gallon plus capacity to carry two 250 imperial gallon ventral tanks.
Weight: Take-off (clean) 32,800lb, Overload 37,480lb (two ventral tanks).
Max speed @Sea Level (clean): 693mph, at 40,000ft 625mph.
Climb to 45,000ft: 10.3 minutes
Ceiling: Service 50,700ft, Absolute 52,000ft.
Armament: Four fixed 30mm ADEN cannon mounted in the outer wing panels.
Production: 50

Designation: F(AW) Mk 5

First Flight: 26th July 1956 (XA641)
Span: 52ft 0in
Length: 56ft 4in
Height: 16ft 0in
Engine: Two Armstrong-Siddeley Sapphire Sa6 Mk 10201 (port) and 10301 (starboard) turbojets offering 8,300lb thrust or two Armstrong-Siddeley Sapphire Sa6 Mk 10701 (port) and 10801 (starboard) turbojets offering 8,300lb thrust.
Fuel Capacity: 995 imperial gallon plus capacity to carry two 250 imperial gallon ventral tanks.
Weight: Take-off (clean) 34,990lb, Overload 39,370lb (two ventral tanks).
Max speed @Sea Level (clean): 695mph, at 40,000ft 608mph.
Climb to 45,000ft: 10.3 minutes
Ceiling: Service 50,100ft, Absolute 51,600ft.
Armament: Four fixed 30mm ADEN cannon mounted in the outer wing panels.
Production: 64

Designation: F(AW) Mk 6

First Flight: 14th December 1956 (XA815)
Span: 52ft 0in
Length: 56ft 3in
Height: 16ft 0in
Engine: Two Armstrong-Siddeley Sapphire Sa6 Mk 10701 (port) and 10801 (starboard) turbojets offering 8,300lb thrust.
Fuel Capacity: 995 imperial gallon plus capacity to carry two 250 imperial gallon ventral tanks.
Weight: Take-off (clean) 35,810lb, Overload 40,600lb (two ventral tanks).
Max speed @Sea Level (clean): 695mph, at 40,000ft 608mph.
Climb to 45,000ft: 10.3 minutes
Ceiling: Service 50,100ft, Absolute 51,600ft.
Armament: Four fixed 30mm ADEN cannon mounted in the outer wing panels.
Production: 33

Designation: F(AW) Mk 7

First Flight: 9th November 1956 (XH704)
Span: 52ft 0in
Length: 56ft 3in
Height: 16ft 0in
Engine: Two Armstrong-Siddeley Sapphire Sa7 Mk 20301 (port) and 20401 (starboard) turbojets offering 11,000lb thrust.
Fuel Capacity: 915 imperial gallon plus capacity to carry two 250 imperial gallon ventral tanks and four 230 imperial gallon underwing tanks.
Weight: Take-off (clean) 35,690lb, Overload 40,270lb (two ventral tanks).
Max speed @Sea Level (clean): 700mph, at 45,000ft 614mph.
Climb to 45,000ft: 6.6 minutes
Ceiling: Service 52,800ft, Absolute 54,100ft.
Armament: Four fixed 30mm ADEN cannon mounted in

the outer wing panels. Plus provision for four
De Havilland Firestreak air-to-air missiles carried on
pylons mounted under the wings, when only one 30mm
cannon was retained in each wing.
Production: 142

Designation: F(AW) Mk 8

First Flight: 9th May 1958 (XH966)
Span: 52ft 0in
Length: 56ft 4in
Height: 16ft 0in
Engine: Two Armstrong-Siddeley Sapphire Sa7R
Mk 20501R (port) and 20601R (starboard) turbojets
offering 11,000lb thrust and with 12% reheat to give
12,300lb thrust above 20,000ft.
Fuel Capacity: 950 imperial gallon plus capacity to carry
two 250 imperial gallon ventral tanks and four 230
imperial gallon underwing tanks.
Weight: Take-off (clean) 37,410lb, Overload 42,510lb
(two ventral tanks).
Max speed @Sea Level (clean): 693mph,
at 35,000ft 607mph.
Climb to 40,000ft: 7.24 minutes
Ceiling: Service 52,000ft, Absolute 54,000ft.
Armament: Four fixed 30mm ADEN cannon mounted in
the outer wing panels. Plus provision for four
De Havilland Firestreak air-to-air missiles carried on
pylons mounted under the wings, when only one 30mm
cannon was retained in each wing.
Production: 47

Designation: F(AW) Mk 9 & 9R

First Flight: 6th May 1959
Span: 52ft 0in
Length: 56ft 4in
Height: 16ft 0in
Engine: Two Armstrong-Siddeley Sapphire Sa7R
Mk 20901R (port) and 21101R (starboard) turbojets
offering 11,000lb thrust and with 12% reheat to give
12,300lb thrust above 20,000ft.
Fuel Capacity: 950 imperial gallon plus capacity to carry
two 250 imperial gallon ventral tanks and four 230
imperial gallon underwing tanks.
Weight: Take-off (clean) 38,100lb, Overload 43,165lb
(two ventral tanks).
Max speed @Sea Level (clean): 693mph,
at 35,000ft 607mph.
Climb to 50,000ft: 9.25 minutes
Ceiling: Service 52,000ft, Absolute 54,000ft.
Armament: Four fixed 30mm ADEN cannon mounted in
the outer wing panels. Plus provision for four
De Havilland Firestreak air-to-air missiles carried on
pylons mounted under the wings, when only one 30mm
cannon was retained in each wing.
Production: 116 (converted from Mk 7s)

Trainer

Designation: T Mk 3

First Flight: 20th August 1956 (WT841)
Span: 52ft 0in
Length: 60ft 3in
Height: 16ft 0in
Engine: Two Armstrong-Siddeley Sapphire Sa6 Mk 11201
(port) and 11301 (starboard) turbojets offering 8,000lb
thrust.

Fuel Capacity: 1,064 imperial gallons plus capacity to
carry two 250 imperial gallon ventral tanks.
Weight: Take-off (clean) 38,000lb, Overload 42,000lb
(two ventral tanks).
Max speed @Sea Level (clean): 650mph,
at 35,000ft 603mph.
Climb to 45,000ft: 22 minutes
Ceiling: Service 46,000ft, Absolute 49,500ft.
Armament: Four fixed 30mm ADEN cannon mounted in
the outer wing panels.
Production: 22

Note
Most published sources quote 59ft 11ins for the length of
the T Mk 3, however the figure shown here is that quoted
by the RAF in their Air Publication for the type; 60ft 3ins
being the 'on the ground' length and 59ft 11ins being the
'in flight' dimensions.

We have refrained from using metric conversions on the
above tables, as the Javelin was built to imperial
specifications and dimensions and any conversion to
metric would require working to many decimal places to be
anywhere near accurate, so we have omitted any mention
of metric at all.

CONVERSION CHART
Imperial
1 nautical mile (nm) = 1. 136 statute miles
1 miles per hour (mph) = 0.88 knots
1 imperial gallon = 4.546 litres

Imperial to Metric
1 foot = 0.3048 metres
1 nautical mile = 1.828 kilometres
1 square foot = 0.0929 square metres
1 ft per minute = 0.005 m/sec
1mph = 1.61 km/h
1lb = 0.454 kg
1 Gallon = 4.546 litres

Appendix II: **JAVELIN SQUADRONS**

NO.3 SQUADRON

Equipped with Javelin: Geilenkirchen West Germany
1/59. (By renumbering 96 Squadron as No 3. 21/1/59).
UK Bases: None.
Foreign Bases: Geilenkirchen, 1/59–12/60.
Disbanded: Geilenkirchen 31/12/60. (Reformed at same
location following month with Canberra B(I).8.)
Variants Operated: F(AW) Mk 4.

NO.5 SQUADRON

Equipped with Javelin: Laarbruch, West Germany
from 1/60 (Meteor NF.11 retained until mid 1960).
UK Bases: None.
Foreign Bases: Laarbruch, 1/59–11/62.
Geilenkirchen, 11/62 to 7/10/65.
Disbanded: Geilenkirchen, 7/10/65. (Reformed same day
with Lightning F.6 at RAF Binbrook).
Variants Operated: F(AW) Mk 5 and 9.

NO.11 SQUADRON

Equipped with Javelin: Geilenkirchen, from 10/59.
UK Bases: None.
Foreign Bases: Geilenkirchen 10/59–12/01/66.
Disbanded: Geilenkirchen 12/01/66. (Reformed Leuchars
01/04/67 with Lightning F.3 and F.6.
Variants Operated: F(AW) Mk 4, 5 and 9.

NO.23 SQUADRON

Equipped with Javelin: Horsham St. Faith from 3/57.
UK Bases: Horsham St. Faith 3/57–5/57,
Coltishall 6/57–3/63, Leuchars 3/63–9/64.
Foreign Bases: None, but short detachments to
West Germany, Middle East and Far East.
Re-equipped: Lightning F.3 Leuchars, from 8/64.
(Javelin F(AW) 9 retained until 10/64).
Variants Operated: F(AW) Mk 4, 5, (single example only)
7 and 9.

NO.25 SQUADRON

Equipped with Javelin: Waterbeach 12/58.
UK Bases: Waterbeach 12/58–10/61,
Leuchars 10/61–11/62.
Foreign Bases: None.
Disbanded: Leuchars 11/62.
Variants Operated: F(AW) Mk 7 and 9.

NO.29 SQUADRON

Equipped with Javelin: Acklington 11/57.
UK Bases: Acklington 11/57–7/58, Leuchars 7/58–2/63.
Foreign Bases: Nicosia 2/63–3/64, (with nine month
detachment to Zambia), Akrotiri 3/64–5/67.
Re-equipped: Lightning F.3 Wattisham, 5/67 having
returned from Akrotiri.
Variants Operated: F(AW) Mk 6 and 9.

NO.33 SQUADRON

Equipped with Javelin: Leeming 4/7/58.
UK Bases: Leeming 7/58, Middleton St. George
9/58–11/62.
Foreign Bases: None.
Disbanded: Middleton St. George 17/11/62.
Variants Operated: F(AW) Mk 7 and 9.

NO.41 SQUADRON

Equipped with Javelin: Coltishall 16/1/58 by
renumbering No.141 Squadron.
UK Bases: Coltishall 1/58–7/58, Wattisham 7/58–12/63
(with detachment to Malta).
Foreign Bases: None.
Disbanded: Wattisham 6/12/63. (Subsequently
re-equipped 9/65 as a Bloodhound unit until 9/70,
reforming again 4/72 with Phantom FGR.2).
Variants Operated: F(AW) Mk 4, 5 and 8.

NO.46 SQUADRON

Equipped with Javelin: Odiham 2/56.
UK Bases: Odiham 2/56–7/59, Waterbeach 7/59–6/61.
Foreign Bases: None.
Disbanded: Waterbeach 30/6/61.
Variants Operated: F(AW) Mk 1, 2 and 6.

NO.60 SQUADRON

Equipped with Javelin: Tengah 7/61.
UK Bases: None.
Foreign Bases: Tengah 7/61–4/68.
Disbanded: Tengah 30/4/68.
Variants Operated: F(AW) Mk 9.

NO.64 SQUADRON

Equipped with Javelin: Duxford 8/58.
UK Bases: Duxford 8/58–7/61, Waterbeach 7/61–8/62,
Binbrook 8/62–4/65.
Foreign Bases: Tengah 4/65–6/67.
Disbanded: Tengah 16/6/67.
Variants Operated: F(AW) Mk 7 and 9.

NO.72 SQUADRON

Equipped with Javelin: Church Fenton 4/59.
UK Bases: Church Fenton 4/59–6/59, Leconfield
7/59–6/61.
Foreign Bases: None.
Disbanded: Leconfield 30/6/61.
Variants Operated: F(AW) Mk 4 and 5.

NO.85 SQUADRON

Equipped with Javelin: Stradishall 30/11/58.
UK Bases: Stradishall 11/58–6/59, West Malling 6/59–9/60, West Raynham 9/60–3/63.
Foreign Bases: None.
Disbanded: West Raynham 31/3/63. (Reformed next day at same location by redesignating the Target Facilities Squadron, operating Meteor F.8s).
Variants Operated: F(AW) Mk 2, 6 and 8.

NO.87 SQUADRON

Equipped with Javelin: Brüggen 8/57.
UK Bases: None.
Foreign Bases: Brüggen 8/57–1/61.
Disbanded: Brüggen 3/1/61.
Variants Operated: F(AW) Mk 1, 4 and 5.

NO.89 SQUADRON

Equipped with Javelin: Stradishall 10/57.
UK Bases: Stradishall 10/57–11/58.
Foreign Bases: None.
Disbanded: Stradishall by renumbering as No.85 Squadron 30/11/58.
Variants Operated: F(AW) Mk 2 and 6.

NO.96 SQUADRON

Equipped with Javelin: Geilenkirchen 9/58.
UK Bases: None.
Foreign Bases: Geilenkirchen.
Disbanded: Geilenkirchen 21/1/59. (Reformed same day, same location as No 3 Squadron).
Variants Operated: F(AW) Mk 4.

NO.141 SQUADRON

Equipped with Javelin: Horsham St. Faith 2/57.
UK Bases: Horsham St. Faith 2/57–5/57. Coltishall 6/57–1/58.
Foreign Bases: None.
Disbanded: Coltishall by renumbering as No.41 Squadron 16/1/58.
Variants Operated: F(AW) Mk 4.

NO.151 SQUADRON

Equipped with Javelin: Leuchars 5/57.
UK Bases: Leuchars 5/57–6/57. Turnhouse 6/57–11/57. Leuchars 11/57–9/61.
Foreign Bases: None.
Disbanded: Leuchars 19/9/61. (1/1/62 Signals Command Development Squadron at Watton, redesignated 151 Squadron operating Lincoln B.2, Hastings C.2 and Canberra amongst other aircraft types!)
Variants Operated: F(AW) Mk 5.

Selected Javelin units

Royal Air Force and other Javelin users

No.1 Guided Weapons Development Squadron
This unit was formed at RAF Valley in June 1957 and was disbanded there in April 1962. Operated F(AW) Mk 7.

No. 228 Operational Conversion Unit (OCU)
This unit operated Javelin F(AW) Mk 5, T Mk 3, Valetta C Mk 1, T Mk 4 and Canberra T Mk 11 at Leeming from June 1957, disbanding in August 1961. Reformed 1st May 1965 at Leuchars using the F(AW) Mk 9, T Mk 3 and Canberra T Mk 11 finally disbanding in December 1966.

Javelin Mobile Conversion Unit (JMCU)
This unit was part of No.228 OCU based at Leeming becoming inactive in January 1958. It restarted operations in July 1958 in readiness for the equipping of units with the F(AW) Mk 7 and it finally disbanded a year later. Types Operated: Javelin F(AW) Mk 5, T Mk 3, Valetta C Mk 1 and T Mk 4.

Javelin Operational Conversion Squadron
This unit was formed from the All-Weather Fighter Combat School (AWFCS) in July 1962. It operated the F(AW) Mk 5 and disbanded in November 1962.

Central Fighter Establishment (CFE)
Major sub-units in the CFE included:
• All-weather Wing, first CFE Javelin user with Mk 1s January 1956.
• All-weather Development Squadron, operated Javelin F(AW) Mks 1, 2, 4, 5, and 7 from October 1956 to August 1960.
• Fighter Command Instrument Rating Squadron. This unit formed at West Rayham in January 1960 moving to Middleton St. George in August 1961 becoming part of 226 OCU (Lightnings) in June 1963. Moved to Leuchars in January 1964 as Javelin IRS and was absorbed into 228 OCU in May 1965. This unit operated up to four T Mk 3s.
• Air Fighting Development Squadron. This unit received its first Javelin F(AW) Mk 8 in October 1959 and later used Mks 1, 6 and 7.

All-weather Fighter Combat School
Operated F(AW) Mk 5s from West Raynham between September 1957 and July 1962 when this became the Javelin Operational Conversion Squadron. Disbanded December 1962.

Government & Research

A&AEE Boscombe Down
This unit tested the following aircraft:
Prototypes and F(AW) Mk 1s WT827, WT830, XA544, XA547, XA548, XA549, XA551, XA554, XA558, XA559, XA561, XA563 and XA623.
F(AW) Mk 2 XD158, XA769, XA770, XA771 and XA778.
T Mk 3 XK577, XH390, XH391 and XH433
F(AW) Mk 4 XA630, XA631, XA720, XA721, XA725 and XA760.
F(AW) Mk 5 XA641, XA649, XA692, XA709 and XA711.
F(AW) Mk 6 XA821, XA831 and XA834.
F(AW) Mk 7 XH704, XH705, XH706, XH708, XH710, XH712, XH713, XH714, XH722, XH746, XH753, XH757, XH780 and XH783. Plus many other known to have been used.
F(AW) Mk 8 XH966, XH967, XH968, XH969, XH970, XH975, XH979 and XJ123.
F(AW) Mk 9 XH759, XH762, XH845, XH848, XH871, XH874, XH891, XH896, XH897, XH964, and XH965.

See Appendix III for details of individual aircraft histories and their use by various company and Government research establishments.

Appendix III: JAVELIN PRODUCTION

GA.5 AND JAVELIN PROTOTYPES

Specification Number: F.4/48.
Contract Number: N/A.
Built by: Gloster Aircraft Company Limited.
Quantity: 7.

WD804 First flight 26th November 1951. Crashed at A&AEE Boscombe Down 29th June 1952 (Beyond Repair).

WD808 First flight 20th August 1952. Crashed at Flax Burton, Bristol 11th June 1953. Pilot (Peter Lawrence) was killed when he ejected too low.

WT827 First flight 7th March 1953, F/R trials. SOC RAE Bedford.

WT830 First flight 14th January 1954. Became Instructional Airframe No.7485M at RAF Halton. Later to Cosford.

WT836 First flight 20th July 1954. Became Instructional Airframe No.7552M at RAF Cosford.

WT841 Gloster P.319 prototype for T Mk 3. First flight 26th August 1956, pilot Jan Zurakowski. Aircraft originally designed to meet OR278 and later Specification T.118D written around it. Prototype assembled by Air Services Training at Hamble. It did not go to A&AEE and all flying trials were conducted by Gloster Aircraft. Had flying slot at SBAC Display 3rd-9th September 1956 flown by Geoff Worrall and was in the static park at SBAC Display 2nd-9th September 1957.

XD158 Gloster P.272 prototype F(AW)Mk 2. First flight 31st October 1955, pilot Wg Cdr 'Dickie' Martin. Revised nose contour, which hinged to left, to accept American APQ43 (AI.22) radar. This aircraft introduced the all-flying tail. It went to A&AEE on 10th May 1956 with metal nose cone but no radar. Did 30hrs flying at A&AEE between May 1956 and February 1957. Became Instructional Airframe No.7592M at RAF Halton.

JAVELIN F(AW) Mk 1

Specification Number: F.4/48.
Contract Number: 6/ACFT/8336 dated 14th July 1952.
Built by: Gloster Aircraft Company Limited.
Quantity: 40. XA544-572, XA618-628. Of these 29 were delivered to Nos.46 and 87 Squadrons, the remaining 11 were either retained by Gloster or issued to various Government establishments for trials.

XA544 Became Instructional Airframe No.7558M at RAF St. Athan.

XA545 This aircraft was used to assist in the development of the all-flying tail used from the F(AW) Mk 4 onwards.

XA546 Used for spinning trials and also test fitted with Gee 3 navigational system. Lost in fatal crash on the 21st October 1954.

XA547 Fitted with four underwing pylons in association with the initial trials of the De Havilland Blue Jay (later Firestreak) missile. Also used by the A&AEE for trials with the ventral ('bosom') tanks. Also used by the All-Weather Development Squadron (CFE) [AWDS], Air Fighting Development Squadron (AFDS) and the College of Aeronautics (Cranfield).

XA548 Retained by Gloster for stalling and spinning trials. Aircraft was fitted with a spin-recovery parachute, tail bumper, a drooped wing leading edge and slats. Aircraft crashed at Ashley on the Isle of Wight while being flown by Sqn Ldr David Dick of A&AEE and after entering a flat spin. Pilot ejected safely, but aircraft totally destroyed.

XA549 No.87 Squadron. Became Instructional Airframe No.7717M at RAF Swanton Morley, gate guardian.

XA550 Retained by Gloster for a variety of trials at Moreton Valence. Became Instructional Airframe 7484M at Weeton.

XA551 Became Instructional Airframe No.7568M at RAF Halton. Scrapped 71 MU in 1964.

XA552 Used as a flying test bed for the De Havilland Gyron Junior engine in association with the Bristol 188 research aircraft. Airframe to Spadeadam 1963.

XA553 Retained by Gloster, never issued to RAF. Became Instructional Airframe No.7470M at RAF Yatesbury (gate guardian, later moved to RAF Stanmore Park). Gate guardian at the Thunder City Entertainment complex, Cape Town Airport, South Africa, extant.

XA554 No.87 Squadron. Became Instructional Airframe No.7662M at RAF St. Athan.

XA555 AWDS (CFE), AFDS and the College of Aeronautics (Cranfield).

XA556 AWDS (CFE), AFDS and the College of Aeronautics (Cranfield).

XA557 Used by Armstrong-Siddeley for flight testing of the Sapphire 100 series engines. Later issued to Bristol-Siddeley for strain gauge test in relation to cracks that had developed in the inlet guide vanes of production aircraft.

XA558 No.87 Squadron.

XA559 No.87 Squadron.

XA560 Used by Armstrong-Siddeley to flight test the Sapphire Sa7 engines fitted with reheat. Became Instructional Airframe No.7619M at RAF Henlow.

XA561 Used by A&AEE for spinning trials, fitted with anti-spin parachute and tail bumper.

XA562 Used by Rolls-Royce for flight testing the Avon RA.24R engine.

XA563 Used by A&AEE for handling trials. Became Instructional Airframe No.7627M at RAF Newton.

XA564 Became Instructional Airframe No.7464M at RAF Locking. Displayed at the RAF Museum, Cosford, extant.

XA565 All-Weather Wing (CFE), No.46 Squadron and No.87 Squadron.

XA566 All-Weather Wing (CFE), No.46 Squadron and No.87 Squadron.

XA567 Became Instructional Airframe No.7551M at RAF Melksham.

XA568 All-Weather Wing (CFE), AWDS (CFE), No.46 Squadron and the College of Aeronautics (Cranfield).

XA569 No.46 Squadron and No.87 Squadron.

XA570 No.46 Squadron. Aircraft crashed whilst with 46 Squadron.

XA571 No.46 Squadron and 87 Squadron. Became Instructional Airframe No.7663M at RAF Halton.

XA572 No.46 Squadron and No.87 Squadron.

XA618 No.46 Squadron and No.87 Squadron.

XA619 No.46 Squadron and No.87 Squadron.

XA620 No.46 Squadron and No.87 Squadron. Became Instructional Airframe No.7723M at RAF Cosford.

XA621 No.46 Squadron and No.87 Squadron.

XA622 No.46 Squadron.

XA623 AWDS (CFE), AFDS, No.46 Squadron, No.87 Squadron and the College of Aeronautics (Cranfield).

XA624 No.46 Squadron and No.87 Squadron.

XA625 No.46 Squadron and No.87 Squadron.

XA626 No.46 Squadron and No.87 Squadron.

XA627 No.46 Squadron and No.87 Squadron. Became Instructional Airframe No.7661M at RAF St. Athan.

XA628 No.46 Squadron and No.87 Squadron. Became Instructional Airframe No.7720M at RAF Melksham.

JAVELIN F(AW) Mk 2

Specification Number: F.4/48.
Contract Number: 6/ACFT/8336 dated 14th July 1952.
Built by: Gloster Aircraft Co., Ltd.
Quantity: 30. XA768-781, XA799-814. Of these 26 went to Nos.46, 85 and 89 Squadron and the remaining four were used for trials.

XA768   No.46 Squadron.
XA769   Used for AI.22 (APQ-43) radar, electrical system and generator development.
XA770   Used by A&AEE for armament trials.
XA771   Used for development trials of the Firestreak missile, when it usually carried just two under the wings.
XA772   No.46 Squadron.
XA773   No.46 Squadron.
XA774   No.89 Squadron and No.85 Squadron.
XA775   No.89 Squadron and No.85 Squadron.
XA776   No.46 Squadron.
XA777   No.46 Squadron.
XA778   Modified to act as the prototype for the F(AW) Mk 7. Later fitted with Sapphire 203/204 engines and used by the A&AEE (1/61) for ASI Pressure Error Correction work until SOC on the 28th March 1969 (replaced by XH897).
XA779   No.89 Squadron.
XA780   No.46 Squadron.
XA781   No.89 Squadron and No.85 Squadron.
XA799   No.89 Squadron and No.85 Squadron.
XA800   No.89 Squadron.
XA801   No.46 Squadron. Became Instructional Airframe No.7739M at RAF Stafford (gate guardian).
XA802   No.46 Squadron.
XA803   No.46 Squadron.
XA804   No.89 Squadron and No.85 Squadron.
XA805   No.46 Squadron.
XA806   AWDS (CFE),
        No.89 Squadron and No.85 Squadron.
XA807   No.46 Squadron.
XA808   AWDS (CFE) and No.46 Squadron.
XA809   AWDS (CFE) and No.46 Squadron.
XA810   No.46 Squadron.
XA811   No.46 Squadron.
XA812   No.46 Squadron.
XA813   No.46 Squadron.
XA814   No.46 Squadron.

JAVELIN T Mk 3

Specification Number: F.4/48.
Contract Number: 6/ACFT/11262 dated 27th September 1954.
Built by: Gloster Aircraft Company Limited.
Quantity: 22. XH390-397, XH432-438, XH443-447, XK577, XM336. All of these were either allocated to squadrons or No.228 OCU, although some were retained for trials relating to types initial introduction.

XH390   A&AEE and No.60 Squadron.
XH391   No.228 OCU.
XH392   No.228 OCU (pre and post-1965), and Javelin Instrument Rating Squadron.
XH393   No.5 Squadron, No.29 Squadron, Fighter Command Instrument Rating Squadron and No.228 OCU.
XH394   No.11 Squadron and No.228 OCU.
XH395   No.25 Squadron, No.29 Squadron, No.46 Squadron, Javelin Instrument Rating Squadron and No.228 OCU (pre and post-1965).
XH396   No.228 OCU, No.29 Squadron and Fighter Command Instrument Rating Squadron.
XH397   No.41 Squadron, Javelin Instrument Rating Squadron, No.226 OCU and No.228 OCU (post 1965).
XH432   No.23 Squadron and Javelin Instrument Rating Squadron.
XH433   No.29 Squadron and Fighter Command Instrument Rating Squadron and A&AEE.

XH434   No.64 Squadron and No.72 Squadron.
XH435   No.64 Squadron, No.85 Squadron, Javelin Instrument Rating Squadron and No.228 OCU.
XH436   No.151 Squadron and Fighter Command Instrument Rating Squadron.
XH437   No.23 Squadron, No.33 Squadron and Fighter Command Instrument Rating Squadron.
XH438   No.72 Squadron and Fighter Command Instrument Rating Squadron.
XH443   No.25 Squadron and Fighter Command Instrument Rating Squadron, No.226 OCU and Javelin Instrument Rating Squadron.
XH444   No.3 Squadron, No.11 Squadron and Fighter Command Instrument Rating Squadron.
XH445   No.60 Squadron, No.64 Squadron and No.87 Squadron.
XH446   No.3 Squadron, No.60 Squadron, No.64 Squadron and No.228 OCU.
XH447   No.5 Squadron.
XK577   This production aircraft was used by A&AEE for trials and is often referred to as being the second prototype even though it was a production airframe.
XM336   Delivered 8/59. 226 OCU crashed 5/11/63.

JAVELIN F(AW) Mk 4

Specification Number: F.4/48.
Contract Number: 6/ACFT/8336 dated 14th July 1952.
Built by: Gloster Aircraft Company Limited.
Quantity: 18. XA629-640, XA644, XA763-767. Most of these saw squadron service but a few were used for trials before or after their service life.

XA629   Retained by Gloster to flight test Kuchemann 'bumps' in an attempt to extend the buffet boundary. RAF Ternhill (gate guardian).
XA630   No.96 Squadron and No.3 Squadron.
XA631   Used by the A&AEE for engineering assessment and radio and operational reliability trials. No.23 Squadron, No.72 Squadron and No.11 Squadron.
XA632   Used by Armstrong-Whitworth for trials in relation to development of the Firestreak missile. Modified to carry four dummy missiles on pylons under wings and then by No.11 Squadron.
XA633   No.96 Squadron, No.3 Squadron and No.11 Squadron.
XA634   Modified by Airwork Ltd and used for in-flight refuelling trials. Became Instructional Airframe No.7641M at RAF Melksham. RAF Leeming gate guardian, extant.
XA635   No.3 Squadron.
XA636   No.141 Squadron, No.41 Squadron and No.87 Squadron.
XA637   No.141 Squadron, No.41 Squadron and No.11 Squadron.
XA638   No.141 Squadron, No.41 Squadron and No.3 Squadron.
XA639   No.141 Squadron, No.41 Squadron and No.87 Squadron.
XA640   No.141 Squadron, No.41 Squadron and No.3 Squadron.
XA644   Used by Gloster in their Mk 7 development programme. Destroyed on 24th August 1956 when it collided with a Hawker Hunter, pilot (Brian Smith) was killed.
XA763   AWDS (CFE), No.96 Squadron, and No.3 Squadron.
XA764   AWDS (CFE), No.3 Squadron and No.11 Squadron.
XA765   No.11 Squadron.
XA766   No.141 Squadron, No.41 Squadron and No.11 Squadron.
XA767   No.141 Squadron, No.41 Squadron and No.11 Squadron.

JAVELIN F(AW) Mk 4

Specification Number: F.4/48.
Contract Number: 6/ACFT/8336 dated 14th July 1952.
Built by: Sir W.G. Armstrong-Whitworth Aircraft Limited.
Quantity: 32. XA720-737, XA749-762. Most of these saw squadron service but a few were used for trials either before or after their service life.

XA720 Used by A&AEE for assessment of handling and No.11 Squadron.
XA721 Used by A&AEE for assessment of handling, No.96 Squadron and No.3 Squadron.
XA722 No.23 Squadron, No.72 Squadron.
XA723 Undertook cold weather trials at the RCAF winterisation test centre at Nameo, Canada. No.11 Squadron.
XA724 Used by Armstrong-Whitworth for aerodynamic flight testing of the Firestreak missile.
XA725 Initially used by Armstrong-Whitworth for aerodynamic flight testing of the Firestreak missile, then moved to A&AEE. No.3 Squadron and No.11 Squadron.
XA726 No.23 Squadron and No.72 Squadron
XA727 No.141 Squadron, No.23 Squadron and No.72 Sqn.
XA728 No.23 Squadron and No.72 Squadron.
XA729 No.23 Squadron and No.72 Squadron.
XA730 All-Weather Combat School, No.23 Squadron and No.72 Squadron.
XA731 No.23 Squadron and No.72 Squadron.
XA732 No.23 Squadron.
XA733 No.23 Squadron, No.72 Squadron and No.87 Sqn.
XA734 No.23 Squadron.
XA735 AWDS (CFE), No.96 Squadron and No.3 Squadron.
XA736 No.23 Squadron and No.72 Squadron.
XA737 No.23 Squadron and No.72 Squadron.
XA749 AWDS (CFE), No.96 Squadron and No.3 Squadron.
XA750 No. 96 Squadron, No.3 Squadron, No.141 Squadron, and No.41 Squadron.
XA751 No.141 Squadron and No.41 Squadron.
XA752 No.23 Squadron and No.72 Squadron.
XA753 No.23 Squadron, No.72 Squadron and All-Weather Combat School.
XA754 No.23 Squadron and No.72 Squadron.
XA755 No.23 Squadron and No.72 Squadron. Became Instructional Airframe No.7725M at RAF St. Athan.
XA756 No.141 Squadron, No.41 Squadron and No.11 Sqn.
XA757 No.141 Squadron, No.41 Squadron and No.87 Sqn.
XA758 No.141 Squadron, No.41 Squadron and No.11 Sqn.
XA759 No.141 Squadron, No.41 Squadron and No.11 Sqn.
XA760 No.141 Squadron, No.41 Squadron and No.11 Squadron. Used by A&AEE as a photographic chase plane after serving with No.11 Squadron.
XA761 No.141 Squadron, No.41 Squadron and No.87 Sqn.
XA762 No.141 Squadron, No.41 Squadron, No.96 Squadron and No.3 Squadron.

JAVELIN F(AW) Mk 5

Specification Number: F.4/48.
Contract Number: 6/ACFT/8336 dated 14th July 1952.
Built by: Gloster Aircraft Company Limited.
Quantity: 20. XA641-643, XA645-661.

XA641 Used by A&AEE for assessment of handling and engineering trials, No.5 Squadron and No.87 Squadron.
XA642 AWDS (CFE).
XA643 AWDS (CFE), No.228 OCU and No.11 Squadron.
XA645 No.87 Squadron.
XA646 Night Fighter Leaders School, No.228 OCU and No.72 Squadron, ended up on dump at Manston (1966).
XA647 No.151 Squadron and No.11 Squadron.
XA648 AWDS (CFE).
XA649 Issued to the RAF Handling Squadron based at Boscombe Down and No.5 Squadron.
XA650 No.151 Squadron and No.11 Squadron.
XA651 No.151 Squadron.
XA652 No.151 Squadron, No.228 OCU. AWFCS.
XA653 No.151 Squadron, No.228 OCU and AWFCS.

XA654 No.23 Squadron and No.72 Squadron.
XA655 No.151 Squadron.
XA656 No.228 OCU and AWFCS.
XA657 AWDS (CFE) and No.5 Sqn.
XA658 No.41 Squadron and No.5 Squadron.
XA659 AWDS (CFE) and No.5 Squadron.
XA660 AWDS (CFE), No.228 OCU and No.11 Squadron.
XA661 No.151 Squadron and No.11 Squadron.

Specification Number: F.4/48.
Contract Numbers: 6/ACFT/8336 dated 14th July 1952. (Qty 38) and 6/ACFT/11329 dated 19th October 1954 (Qty 6).
Built by: Sir W.G. Armstrong-Whitworth Aircraft Limited.
Quantity: 44. XA662-667, XA688-719, XH687-692.

XA662 No.228 OCU.
XA663 No.228 OCU, AWFCS, No.11 Squadron and No.5 Squadron.
XA664 No.228 OCU, AWFCS and No.5 Squadron.
XA665 AWFCS, No.228 OCU and No.11 Squadron.
XA666 No.228 OCU, No.41 Squadron and No.5 Squadron.
XA667 No. 228 OCU, No.41 Squadron, No.72 Squadron and No.11 Squadron.
XA688 AWFCS, No.228 OCU and No.151 Squadron.
XA689 No.228 OCU, AWFCS, No.11 Squadron and No.5 Squadron.
XA690 No.228 OCU and No.11 Squadron.
XA691 No.228 OCU, AWFCS and No.11 Squadron.
XA692 Used by the Institute of Aviation Medicine at RAE Farnborough.
XA693 No.228 OCU.
XA694 No.228 OCU, No.151 Squadron and No.11 Squadron.
XA695 No.228 OCU and No.11 Squadron.
XA696 No.11 Squadron and AWFCS.
XA697 AWFCS and No.5 Squadron.
XA698 No.228 OCU.
XA699 No.151 Squadron and No.5 Squadron. Became Instructional Airframe No.7809M at RAF Locking (and later RAF Cosford). Displayed at the Midland Air Museum, extant.
XA700 No.228 OCU and AWFCS.
XA701 No.228 OCU and AWFCS. Became Instructional Airframe No.77765M at RAF Bicester.
XA702 No.228 OCU and AWFCS.
XA703 No.228 OCU, No.41 Squadron, No.72 Squadron and AWFCS.
XA704 CFE and No.5 Squadron.
XA705 AWFCS and No.5 Squadron.
XA706 No.228 OCU. Became Instructional Airframe No.7649M at RAF Leeming (gate guardian).
XA707 No.41 Squadron and No.5 Squadron.
XA708 No.151 Squadron.
XA709 Used by A&AEE for G.90 camera trials. No.5 Squadron.
XA710 No.151 Squadron.
XA711 Used by Armstrong-Whitworth for Sapphire 6 development and for the testing of vortex generators. Later used by A&AEE for ammunition temperature tests, gas chamber icing tests, as a chase plane and on Pitot Error Correction work. It was later used for gravel-arrested landing trials.
XA712 No.151 Squadron.
XA713 No.151 Squadron.
XA714 No.228 OCU, No.151 Squadron and No.11 Squadron.
XA715 No.151 Squadron and No.11 Squadron.
XA716 No.228 OCU and No.11 Squadron.
XA717 No.151 Squadron and No.11 Squadron.
XA718 No.228 OCU, AWFCS and No.5 Squadron.
XA719 No.228 OCU.
XH687 No.151 Squadron.
XH688 No.151 Squadron, No.228 OCU and No.11 Squadron.
XH689 No.151 Squadron, No.228 OCU and No.11 Squadron.
XH690 No.151 Squadron and No.5 Squadron.
XH691 No.228 OCU.
XH692 No.228 OCU.

JAVELIN F(AW) Mk 6

Specification Number: F.4/48.
Contract Numbers: 6/ACFT/8336 dated 14th July 1952 (Qty 22) and 6/ACFT/11329 dated 19th October 1954 (Qty 11).
Built by: Gloster Aircraft Company Limited.
Quantity: 33. XA815-836, XH693-703.

XA815   No.89 Squadron and No.85 Squadron.
XA816   No.89 Squadron and No.85 Squadron. After its service life this aircraft was used by the RAF Fire Fighting Training School at Catterick.
XA817   No.29 Squadron.
XA818   No.29 Squadron.
XA819   No.29 Squadron.
XA820   No.89 Squadron and No.85 Squadron. Became Instruction Airframe No.7752M at RAF Dishforth. RAF Acklington gate guardian.
XA821   Used by A&AEE for target towing experiments, No.29 Squadron. Became Instructional Airframe No.7749M at RAF Hartlebury (gate guardian).
XA822   No.29 Squadron.
XA823   No.29 Squadron.
XA824   No.29 Squadron.
XA825   No.29 Squadron.
XA826   No.29 Squadron.
XA827   No.29 Squadron.
XA828   No.29 Squadron.
XA829   No.29 Squadron.
XA830   No.46 Squadron, No.89 Squadron and No.85 Squadron. After its service life this aircraft was used by the RAF Fire Fighting Training School at Catterick.
XA831   Used by RAE Farnborough for rain erosion trials and later used by the Institute of Aviation Medicine.
XA832   No.89 Squadron, No.85 Squadron and AFDS (CFE).
XA833   No.23 MU.
XA834   Used by A&AEE for handling assessment and fuel consumption tests.
XA835   No.29 Squadron.
XA836   No.89 Squadron, No.85 Squadron and No.29 Squadron.
XH693   No.89 Squadron and No. 85 Squadron.
XH694   No.89 Squadron and No. 85 Squadron.
XH695   No.89 Squadron and No. 85 Squadron.
XH696   No.89 Squadron and No. 85 Squadron.
XH697   No.29 Squadron.
XH698   No.29 Squadron.
XH699   No.29 Squadron.
XH700   No.29 Squadron.
XH701   No.29 Squadron.
XH702   No.46 Squadron, No.89 Squadron, No.85 Squadron and AFDS (CFE).
XH703   No.29 Squadron.

JAVELIN F(AW) Mk 7

Specification Number: F.4/48.
Contract Number: 6/ACFT/11329 dated 19th October 1954.
Built by: Gloster Aircraft Company Limited.
Quantity: 85. XH704-725, XH746-784, XH900-912, XH955-965.

XH704   Initially used by A&AEE for performance trials, then used by Gloster for handling assessment.
XH705   Used by A&AEE for Firestreak production missile installation trials and later used as a photographic chase plane in relation to ejector seat trials.
XH706   Used by A&AEE for handling assessment and autostabiliser trials.
XH707   Used by Armstrong-Siddeley for Sapphire Sa7R engine evaluation with different forms of reheat in relation to the Mk 8 programme. Converted to F(AW) Mk 9.
XH708   Used by Gloster, the A&AEE for engineering trials. Converted to F(AW) Mk 9.
XH709   Converted to F(AW) Mk 9.

XH710   Used by A&AEE for radio installation, performance and operational reliability trials. Later used by them as a pace plane. Became Instructional Airframe No.7748M at RAF Melksham.
XH711   Used by A&AEE for autopilot and instrument evaluation trials. Later used for tropical trials of Firestreak in Cyprus. Converted to F(AW) Mk 9.
XH712   Used for handling trials. Converted to F(AW) Mk 9.
XH713   Used for tropical trials in Bahrain. Converted to F(AW) Mk 9.
XH714   Used by A&AEE for operational reliability trials. Destroyed 26th February 1958 when pilot was inadvertently ejected, the navigator deliberately ejected but both were killed due to a fault in the ejection process which damaged both seats.
XH715   No.33 Squadron. Converted to F(AW) Mk 9.
XH716   No.33 Squadron. Converted to F(AW) Mk 9.
XH717   No.64 Squadron. Converted to F(AW) Mk 9.
XH718   No.33 Squadron.
XH719   No.33 Squadron. Converted to F(AW) Mk 9.
XH720   No.33 Squadron.
XH721   No.33 Squadron. Converted to F(AW) Mk 9.
XH722   Used by Gloster and A&AEE for tropical trials in Libya. Converted to F(AW) Mk 9.
XH723   No.64 Squadron. Converted to F(AW) Mk 9.
XH724   No.64 Squadron. Converted to F(AW) Mk 9.
XH725   No.64 Squadron. Converted to F(AW) Mk 9.
XH746   Used for Sapphire Sa7R with reheat trials in relation to Mk 8 programme. Later used by Bristol-Siddeley for strain gauging, booster pump tests, relighting and reheat relighting trials. Later used by A&AEE for noise attenuation trials.
XH747   AWDS and No.64 Squadron. Converted to F(AW) Mk 9.
XH748   AWDS (CFE), No.33 Sqn.
XH749   AWDS, AFDS and No.64 Squadron. Converted to F(AW) Mk 9.
XH750   No.33 Squadron.
XH751   No.33 Squadron. Converted to F(AW) Mk 9.
XH752   AWDS and No.64 Squadron. Converted to F(AW) Mk 9.
XH753   Various trials with Armstrong-Siddeley and A&AEE. Converted to F(AW) Mk 9.
XH754   Used by Gloster for handling and autopilot trials at A&AEE. Later used by RAE Farnborough for various trials in the Structural Engineering Flight. Undertook much of this work away from the UK and on return was used for parachute trials on Larkhill Ranges. Often referred to as the Javelin Mk 7.5.
XH755   AWDS. Converted to F(AW) Mk 9.
XH756   Used for missile trials and by the AWDS (CFE). Converted to F(AW) Mk 9.
XH757   Used by A&AEE for tropical trials at Kano, Nigeria. Converted to F(AW) Mk 9.
XH758   AWDS and No.23 Squadron. Converted to F(AW) Mk 9.
XH759   Used by A&AEE for handling trials. Converted to F(AW) Mk 9.
XH760   Converted to F(AW) Mk 9.
XH761   Converted to F(AW) Mk 9.
XH762   Converted to F(AW) Mk 9.
XH763   Converted to F(AW) Mk 9.
XH764   Converted to F(AW) Mk 9.
XH765   Converted to F(AW) Mk 9.
XH766   Converted to F(AW) Mk 9.
XH767   Converted to F(AW) Mk 9.
XH768   Converted to F(AW) Mk 9.
XH769   Converted to F(AW) Mk 9.
XH770   Converted to F(AW) Mk 9.
XH771   Converted to F(AW) Mk 9.
XH772   Converted to F(AW) Mk 9.
XH773   Converted to F(AW) Mk 9.
XH774   No.23 Squadron. Converted to F(AW) Mk 9.
XH775   No.23 Squadron.
XH776   Converted to F(AW) Mk 9.
XH777   No.23 Squadron. Converted to F(AW) Mk 9.

XH778   No.23 Squadron. Converted to F(AW) Mk 9.
XH779   No.23 Squadron. Converted to F(AW) Mk 9.
XH780   Used by Flight Refuelling Ltd for in-flight refuelling trials. Converted to F(AW) Mk 9.
XH781   No.23 Squadron.
XH782   No.1 Guided Weapons Test Squadron (GWTS). Became Instructional Airframe No.7797M at RAF Halton.
XH783   Used by A&AEE and No.1 GWTS (Valley). Became Instructional Airframe No.7798M at RAF Halton.
XH784   No.1 GWTS. Became Instructional Airframe No.7799M at RAF Halton.
XH900   No.1 GWTS. Became Instructional Airframe No.7811M at RAF St. Athan. RAF Swinderby gate guardian.
XH901   No.1 GWTS. Became Instructional Airframe No.7800M at RAF Weeton.
XH902   No.1 GWTS. Became Instructional Airframe No.7801M at RAF Weeton.
XH903   No.23 Squadron. Converted to F(AW) Mk 9.
XH904   No.23 Squadron. Converted to F(AW) Mk 9.
XH905   No.25 Squadron. Converted to F(AW) Mk 9.
XH906   No.25 Squadron. Converted to F(AW) Mk 9.
XH907   No.25 Squadron. Converted to F(AW) Mk 9.
XH908   No.25 Squadron. Converted to F(AW) Mk 9.
XH909   No.25 Squadron. Converted to F(AW) Mk 9.
XH910   No.25 Squadron. Converted to F(AW) Mk 9.
XH911   No.25 Squadron. Converted to F(AW) Mk 9.
XH912   No.25 Squadron. Converted to F(AW) Mk 9.
XH955   No.25 Squadron, No.23 Squadron. Converted to F(AW) Mk 9.
XH956   No.25 Squadron, No.23 Squadron. Converted to F(AW) Mk 9.
XH957   No.25 Squadron. Converted to F(AW) Mk 9.
XH958   No.23 Squadron. Converted to F(AW) Mk 9.
XH959   No.25 Squadron. Converted to F(AW) Mk 9.
XH960   No.23 Squadron. Converted to F(AW) Mk 9.
XH961   No.25 Squadron. Converted to F(AW) Mk 9.
XH962   No.23 Squadron. Converted to F(AW) Mk 9.
XH963   No.23 Squadron. Converted to F(AW) Mk 9.
XH964   No.23 Squadron. Converted to F(AW) Mk 9.
XH965   Converted to F(AW) Mk 9.

Specification Number: F.4/48.
Contract Number: 6/ACFT/11329 dated 19th October 1954.
Built by: Sir W.G. Armstrong-Whitworth Aircraft Ltd.
Quantity: 57. [XH785-795, XH833-849, XH871-899].

XH785   No.64 Squadron. Converted to F(AW) Mk 9.
XH786   No.33 Squadron.
XH787   No.64 Squadron. Converted to F(AW) Mk 9.
XH788   No.64 Squadron. Converted to F(AW) Mk 9.
XH789   No.64 Squadron.
XH790   AWDS and No.33 Squadron. Became Instructional Airframe No.7808M at RAF Yatesbury.
XH791   No.64 Squadron. Converted to F(AW) Mk 9.
XH792   AWDS (CFE) and No.64 Squadron. Converted to F(AW) Mk 9.
XH793   Converted to F(AW) Mk 9.
XH794   No.64 Squadron. Converted to F(AW) Mk 9.
XH795   No.33 Squadron.
XH833   No.33 Squadron. Converted to F(AW) Mk 9.
XH834   No.64 Squadron. Converted to F(AW) Mk 9.
XH835   No.33 Squadron. Converted to F(AW) Mk 9.
XH836   No.33 Squadron. Converted to F(AW) Mk 9.
XH837   No.33 Squadron. Forward fuselage only displayed at the Caernarfon Airworld Museum, Caernarfon Aerodrome, Gwynedd, Wales, extant.
XH838   No.33 Squadron.
XH839   No.33 Squadron. Converted to F(AW) Mk 9.
XH840   No.64 Squadron. Converted to F(AW) Mk 9.
XH841   No.64 Squadron. Converted to F(AW) Mk 9.
XH842   No.64 Squadron. Converted to F(AW) Mk 9.
XH843   Converted to F(AW) Mk 9.
XH844   Converted to F(AW) Mk 9.

XH845   Converted to F(AW) Mk 9.
XH846   No.64 Squadron. Converted to F(AW) Mk 9.
XH847   Converted to F(AW) Mk 9.
XH848   Converted to F(AW) Mk 9.
XH849   No.23 Squadron.
XH871   Converted to F(AW) Mk 9.
XH872   Converted to F(AW) Mk 9.
XH873   Converted to F(AW) Mk 9.
XH874   Converted to F(AW) Mk 9.
XH875   Converted to F(AW) Mk 9.
XH876   Converted to F(AW) Mk 9.
XH877   Converted to F(AW) Mk 9.
XH878   Converted to F(AW) Mk 9.
XH879   Converted to F(AW) Mk 9.
XH880   Converted to F(AW) Mk 9.
XH881   Converted to F(AW) Mk 9.
XH882   Converted to F(AW) Mk 9.
XH883   Converted to F(AW) Mk 9.
XH884   Converted to F(AW) Mk 9.
XH885   Converted to F(AW) Mk 9.
XH886   Converted to F(AW) Mk 9.
XH887   Converted to F(AW) Mk 9.
XH888   Converted to F(AW) Mk 9.
XH889   Converted to F(AW) Mk 9.
XH890   Converted to F(AW) Mk 9.
XH891   Converted to F(AW) Mk 9.
XH892   Converted to F(AW) Mk 9.
XH893   Converted to F(AW) Mk 9.
XH894   Converted to F(AW) Mk 9.
XH895   Converted to F(AW) Mk 9.
XH896   Converted to F(AW) Mk 9.
XH897   No.25 Squadron. Converted to F(AW) Mk 9.
XH898   No.25 Squadron. Converted to F(AW) Mk 9.
XH899   No.25 Squadron. Converted to F(AW) Mk 9.

JAVELIN F(AW) Mk 8

Specification Number: F.4/48.
Contract Number: 6/ACFT/11329 dated 19th October 1954.
Built by: Gloster Aircraft Company Limited.
Quantity: 47. XH966-993, *XJ112, XJ113-130, XJ165.

XH966   Used by Gloster for flight trials and later used by A&AEE for high altitude and gun armament trials, No.41 Squadron.
XH967   Used by A&AEE for radio and armament trials, No.41 Squadron.
XH968   Used by Gloster for development work and at A&AEE for armament and radio trials. No.41 Squadron and No.85 Squadron.
XH969   Used by A Squadron, A&AEE for handling and autopilot trials and No.41 Squadron.
XH970   Used by Gloster for handling assessment. Then used by A&AEE for autopilot trials and later for jettison and braking trials.
XH971   No.41 Squadron.
XH972   CFE and AFDS. Became Instructional Airframe No.7834M at RAF Newton.
XH973   No.41 Squadron.
XH974   No.41 Squadron.
XH975   CFE and AFDS.
XH976   CFE and AFDS.
XH977   No.41 Squadron.
XH978   No.41 Squadron.
XH979   CFE and AFDS.
XH980   No.41 Squadron. Became Instructional Airframe No.7867M at RAF Stafford, later moved to RAF West Raynham.
XH981   No.41 Squadron.
XH982   No.41 Squadron.
XH983   No.41 Squadron.
XH984   No.41 Squadron.
XH985   No.41 Squadron.
XH986   No.41 Squadron. Became Instructional Airframe No.7842M at RAF Swanton Morley.

XH987   No.41 Squadron.
XH988   No.41 Squadron.
XH989   No.41 Squadron.
XH990   No.41 Squadron.
XH991   No.85 Squadron. Became Instructional Airframe
         No.7831M at RAF Henlow, then RAF Cranwell.
XH992   No.85 Squadron. Became Instructional Airframe
         No.7829M at RAF Cosford. Displayed at the Newark
         Air Museum, extant.
XH993   No.41 Squadron and No.85 Squadron.
XJ112*   This aircraft was never completed, as it was taken
         from the production line and used for static tests.
XJ113   No.41 Squadron.
XJ114   No.85 Squadron.
XJ115   No.85 Squadron.
XJ116   No.85 Squadron. Became Instructional Airframe
         No.7832M at RAF Newton.
XJ117   No.85 Squadron. Became Instructional Airframe
         No.7833M at RAF Newton.
XJ118   No.85 Squadron.
XJ119   No.85 Squadron.
XJ120   No.85 Squadron.
XJ121   No.85 Squadron.
XJ122   No.85 Squadron.
XJ123   No.85 Squadron.
XJ124   No.85 Squadron.
XJ125   Used by Armstrong-Siddeley for Sapphire Sa7R
         development trials.
XJ126   No.85 Squadron.
XJ127   No.41 Squadron.
XJ128   No.85 Squadron.
XJ129   No.41 Squadron.
XJ130   No.41 Squadron.
XJ165   No.41 Squadron.

JAVELIN F(AW) Mk 9

Specification Number: F.4/48.
Contract Number: N/A
Built by: Gloster Aircraft Company Limited
Quantity: 116. All converted from F(AW) Mk 7s.
XH707-709, XH711-713, XH715-717, XH721-725,
XH747, XH749, XH751-753, XH755-774,
XH776-780, XH785, XH787-788, XH791-794,
XH833-836, XH839-848, XH871-899, XH903-912,
XH955-965. See Note.
XH707   No.23 Squadron, No.64 Squadron and No.60
         Squadron.
XH708   No.64 Squadron.
XH709   No.64 Squadron.
XH711   No.64 Squadron and No.29 Squadron.
XH712   No.23 Squadron, No.29 Squadron.
XH713   No.33 Squadron, No.5 Squadron and No.228 OCU.
XH715   No.33 Squadron, No.5 Squadron and No.228 OCU.
XH716   No.25 Squadron, No.11 Squadron and No.228 OCU.
XH717   No.60 Squadron.
XH719   No.60 Squadron.
XH721   No.60 Squadron.
XH722   No.60 Squadron.
XH723   No.29 Squadron.
XH724   No.60 Squadron.
XH725   No.29 Squadron and No.60 Squadron.
XH747   No.60 Squadron.
XH749   No.29 Squadron and No.60 Squadron.
XH751   No.60 Squadron.
XH752   No.11 Squadron, No.5 Squadron and No.29 Sqn.
XH753   No.11 Squadron and No.5 Squadron.
XH755   No.33 Squadron.
XH756   No.33 Squadron, No.11 Squadron and No.5 Sqn.
XH757   No.33 Squadron and No.5 Squadron.
XH758   No.33 Squadron and No.5 Squadron.
XH759   Used by A&AEE and Gloster for radio and
         roll-damping trials, No.64 Squadron, No.60 Sqn.
XH760   No.25 Squadron and No.11 Squadron. Became
         Instructional Airframe No.7892M at RAF Cranwell.

XH761   Written off 5/60 whilst with Glosters.
XH762   No.64 Squadron, No.5 Squadron and No.29 Sqn.
XH763   No.23 Squadron, No.64 Squadron and No.60 Sqn.
XH764   No.64 Squadron and No.29 Squadron. RAF Manston
         gate guardian.
XH765   No.64 Squadron.
XH766   No.64 Squadron and No.60 Squadron.
XH767   No.25 Squadron and No.11 Squadron and No.228
         OCU. Became Instructional Airframe No.7955M at
         RAF Worcester. Displayed at the Yorkshire Air
         Museum, Elvington, Yorkshire, extant.
XH768   No.25 Squadron, No.11 Squadron and No.29
         Squadron. Became Instructional Airframe No.7929M
         at RAF Cranwell. Displayed at the Museo
         dell'Aviazione, Cerbaiola, near Rimini, Italy and
         marked as 'XH707', extant.
XH769   No.25 Squadron, No.11 Squadron and No.60 Sqn.
XH770   No.25 Squadron, No.11 Squadron and No.64 Sqn.
XH771   No.25 Squadron and No.11 Squadron.
XH772   No.25 Squadron, No.11 Squadron and No.228 OCU.
XH773   No.33 Squadron, No.5 Squadron and No.11 Sqn.
XH774   No.29 Squadron.
XH776   No.25 Squadron, No.11 Squadron and No.29 Sqn.
XH777   No.29 Squadron and No.60 Squadron.
XH778   No.29 Squadron.
XH779   No.29 Squadron and No.60 Squadron.
XH780   Used by Flight Refuelling Ltd for in-flight refuelling
         trials, No.33 Squadron, No.5 Squadron and No.11
         Squadron.
XH785   No.60 Squadron.
XH787   No.60 Squadron.
XH788   No.60 Squadron.
XH791   Crashed 5/8/61 East Pakistan on delivery to No.60 Sqn.
XH792   No.29 Squadron and No.60 Squadron.
XH793   No.23 Squadron, No.64 Squadron and No.60 Sqn.
XH794   No.33 Squadron.
XH833   No.60 Squadron.
XH834   No.29 Squadron and No.64 Squadron.
XH835   No.60 Squadron.
XH836   No.60 Squadron.
XH839   No.60 Squadron.
XH840   Became Instructional Airframe No.7740M at RAF
         Luqa, after damage en route to No.60 Squadron.
XH841   No.60 Squadron.
XH842   No.60 Squadron.
XH843   No.64 Squadron and No.60 Squadron.
XH844   No.64 Squadron.
XH845   A&AEE and No.23 Squadron.
XH846   No.60 Squadron.
XH847   No.23 Squadron and No.29 Squadron.
XH848   No.23 Squadron, No.64 Squadron and No.29
         Squadron. Became Instructional Airframe No.7975M
         at RAF Bicester.
XH871   A&AEE and No.64 Squadron.
XH872   No.64 Squadron and No.60 Squadron.
XH873   No.64 Squadron, No.23 Squadron and No.29 Sqn.
XH874   No.64 Squadron and A&AEE.
XH875   No.64 Squadron.
XH876   No.64 Squadron and No.60 Squadron.
XH877   No.64 Squadron, No.60 Squadron.
XH878   No.64 Squadron.
XH879   No.64 Squadron and No.60 Squadron.
XH880   No.25 Squadron and No.11 Squadron.
XH881   No.25 Squadron and No.11 Squadron.
XH882   No.25 Squadron and No.11 Squadron.
XH883   No.25 Squadron, No.11 Squadron and No.228 OCU.
XH884   No.25 Squadron, No.11 Squadron and No.29 Squadron.
XH885   No.23 Squadron, No.60 Squadron and No.64 Sqn.
XH886   No.23 Squadron, No.64 Squadron and No.29 Sqn.
XH887   No.23 Squadron, No.64 Squadron and No.60 Sqn.
XH888   No.23 Squadron, No.64 Squadron and No.29 Sqn.
XH889   No.23 Squadron, No.64 Squadron and No.29 Sqn.
XH890   No.23 Squadron and No.29 Squadron.
XH891   No.23 Squadron, No.64 Squadron and No.29
         Squadron. Used by A&AEE for camera trials.

XH892 No.23 Squadron, No.64 Squadron and No.29 Squadron. Became Instructional Airframe No.7892M (Cosford). Displayed at the Norfolk & Suffolk Aviation Museum, Flixton, Suffolk, extant.

XH893 No.23 Squadron, No.64 Squadron and No.60 Sqn.

XH894 No.23 Squadron and No.29 Squadron.

XH895 No.23 Squadron, No.64 Squadron and No.60 Sqn.

XH896 No.64 Squadron, A&AEE and No.60 Squadron.

XH897 No.33 Squadron, No.5 Squadron. Bristol-Siddeley (from 7/65) and then used by A&AEE (from 1/4/68) for ASI Pressure Error Correction work. Also used as photographic chase plane including Concorde 01 and Tornado Multi Role Combat Aircraft. Retired and flown to Imperial War Museum, Duxford 24/1/75, extant.

XH898 No.25 Squadron, No.11 Squadron and No.228 OCU.

XH899 No.23 Squadron and No.29 Squadron.

XH903 No.33 Squadron and No.5 Squadron. Became Instructional Airframe No.7938M at RAF Innsworth (gate guardian). Stored at the Jet Age Museum, RAF Fairford, Gloucestershire, extant.

XH904 No.29 Squadron, No.33 Squadron and No.5 Sqn.

XH905 No.33 Squadron and No.5 Squadron and No.228 OCU.

XH906 No.25 Squadron.

XH907 No.33 Squadron and No.5 Sqn and No.228 OCU.

XH908 No.64 Squadron and No.60 Squadron.

XH909 No.25 Squadron and No.11 Sqn and No.228 OCU.

XH910 No.29 Squadron and No.60 Squadron.

XH911 No.33 Squadron and No.5 Squadron.

XH912 No.33 Squadron, No.5 Squadron and No.228 OCU.

XH955 No.60 Squadron.

XH956 No.33 Squadron, No.29 Squadron and No.60 Sqn.

XH957 No.33 Squadron and No.5 Squadron.

XH958 No.33 Squadron, No.5 Squadron, No.29 Squadron and No.228 OCU.

XH959 No.64 Squadron, No.29 Squadron and No.60 Sqn.

XH960 No.29 Squadron and No.60 Squadron.

XH961 No.29 Squadron, No.64 Squadron and No.60 Sqn.

XH962 Used for in-flight refuelling trials and No.29 Sqn.

XH963 No.29 Squadron and No.64 Squadron.

XH964 Used by A&AEE for radio trials, No.29 Squadron and No.60 Squadron.

XH965 Used by Flight Refuelling Ltd for in-flight refuelling trials/ Later used by A&AEE and RAE (Bedford) for ground trials.

Note: The conversion of the above was broken down into three types; Mk 9, 9(F/R) and 9R. The F/R designation denotes that these machines were modified to accept the in-flight refuelling boom, while F(AW) Mk 9Rs were fitted with the in-flight refuelling system and could carry four 230 imperial gallon drop tanks on the underwing pylons.

Converted to F(AW) Mk 9
XH711, XH713, XH715, XH716-719, XH721-725, XH747, XH749, XH751-753, XH755-758, XH760-761, XH767-774, XH776-779, XH785, XH787-788, XH791-792, XH794, XH833-836, XH839-842, XH846, XH880-884, XH897-898, XH903-907, XH903-907, XH909-912, XH956-958, XH960, XH962-964.

Converted to F(AW) Mk 9(F/R)
XH780, XH844, XH875 & XH878.

Converted to F(AW) Mk 9(F/R), then into F(AW) Mk 9R
XH707-709, XH712, XH759, XH762-766, XH793, XH843, XH845, XH847-848, XH871-874, XH876-877, XH879, XH885-896, XH899, XH908, XH955, XH959, XH961 & XH965.

Total Production: 435 new airframes including 116 conversions and prototypes.

Note: Serial numbers emboldened were still extant at February 2005.

A No.41 Squadron (or ex-No.41 Squadron?) Javelin, and presumably therefore a F(AW) Mk 8. via Martyn Chorlton

*Javelins in production in
1955 and final assembly in
1960, both views being
taken at Gloster's
Hucclecote factory.*
Derek N James

XH766, 'E' of No.64 Squadron having taken off from Khormaksar to start the next leg of its flight to Tengah in 1963. Ray Deacon

GLOSSARY AND BIBLIOGRAPHY

A&AEE	Aeroplane & Armament Experimental Establishment
AC	Alternating Current
ACM	Air-Chief Marshal (RAF)
AC1	Aircraftsman 1st Class (RAF)
AC2	Aircraftsman 2nd Class (RAF)
AFB	Air Force Base
AFC	Air Force Cross
AFDS	Air Fighting Development Squadron
AI	Airborne Interception (radar)
Air Cdre	Air Commodore (RAF)
Air Mshl	Air Marshal (RAF)
AOC	Air Officer Commanding
AVM	Air Vice Marshal (RAF)
AWA	Armstrong-Whitworth Aircraft Ltd
AWDS	All-Weather Development Squadron
AWFCS	All-Weather Fighter Combat School
BS	British Standard
CA	Controller Aircraft
Capt	Captain
CFE	Central Fighter Establishment
CGS	Central Gunnery School
CinC	Commander-in-Chief
CO	Commanding Officer
CofG	Centre of Gravity
DC	Direct Current
DF	Direction Finding
DFC	Distinguished Flying Cross
DFM	Distinguished Flying Medal
DH	De Havilland
DSC	Distinguished Service Cross
DSO	Distinguished Service Order
DTD	Directorate of Technical Development
FAA	Fleet Air Arm
F(AW)	Fighter (All-Weather)
FEAF	Far East Air Force
Fg Off	Flying Officer (RAF)
Flt	Flight
Flt Sgt	Flight Sergeant
Flt Lt	Flight Lieutenant
FS	Federal Standard
ft	Foot
g	Acceleration of free fall due to gravity
GAC	Gloster Aircraft Co., Ltd
GHQ	General Headquarters
Gp Capt	Group Captain (RAF)
GWTS	Guided Weapons Test Squadron
H.E.	High Explosive
HMS	His/Her Majesty's Ship
HQ	Headquarters
HSA	Hawker-Siddeley Aviation Ltd
IAM	Institute of Aviation Medicine
IFF	Identification Friend or Foe
ILS	Instrument Landing System
in.	Inch
JMCU	Javelin Mobile Conversion Unit
JMTU	Javelin Mobile Training Unit
kg	Kilogram
km	Kilometre
km/h	Kilometres Per Hour
knot (kt)	Unit of speed of 1nm per hour (approx 1.15mph or 1.85km/h)
LAC	Leading Aircraftsman
lb	Pound
lt	Litre
Maj	Major
MAP	Ministry of Aircraft Production
MEAF	Middle East Air Force
Mk	Mark
MOD	Ministry of Defence
MofA	Ministry of Aviation
MofS	Ministry of Supply
mph	Miles Per Hour
MU	Maintenance Unit (RAF)
NATO	North Atlantic Treaty Organisation
No.	Number

NCO	Non-commissioned Officer
OC	Officer Commanding
OCU	Operational Conversion Unit
OR	Operational Requirement
OTU	Operational Training Unit
Plt Off	Pilot Officer (RAF)
RAE	Royal Aircraft Establishment
RAF	Royal Air Force
SAC	Senior Aircraftsman (RAF)
SBAC	Society of British Aircraft Constructors
Sgt	Sergeant
SOC	Struck Off Charge
SoTT	School of Technical Training
Sqn	Squadron
Sqn Ldr	Squadron Leader (RAF)
T	Trainer
TAF	Tactical Air Force
UHF	Ultra-High Frequency
UK	United Kingdom
USAF	United States Air Force
USS	United States Ship
VC	Victoria Cross
Wg Cdr	Wing Commander (RAF)
W/O	Warrant Officer
W/T	Wireless Telegraphy

British Secret Projects - Jet Fighters since 1950
Tony Buttler, Published by Midland Publishing ©2000
ISBN: 1-85780-095-8

Gloster Aircraft since 1917
Derek James, Published by Putnam ©1971
ISBN: 0-370-00084-6

Gloster Javelin Postwar Military Aircraft No.1
Maurice Allward, Published by Ian Allan ©1983
ISBN: 0-7110-1323-3

Gloster Javelin, Warpaint Series No.17
Tony Buttler AMRAes, Published by Hall Park Books ©1999

Gloster Javelin 1-6, Profile No.179
J J Partridge, Profile Publications Ltd ©1967

Testing Colours
Adrian Balch, Published by Airlife ©1993
ISBN 1-85310-349-7

The Service History of the Gloster Javelin Vol. 1 Mks 1 to 6
Roger Lindsay

The Service History of the Gloster Javelin Vol. 2 Mks 7 to 9
Roger Lindsay

Official Publications
Air Ministry/MoD Air Publications
F(AW) Mk 1 - Air Publication 4491A
F(AW) Mk 2 - Air Publication 4491B
T Mk 3 - Air Publication 4491C
F(AW) Mk 4 - Air Publication 4491D
F(AW) Mk 5 - Air Publication 4491E
F(AW) Mk 6 - Air Publication 4491F
F(AW) Mk 7 - Air Publication 4491G
F(AW) Mk 8 - Air Publication 4491H
F(AW) Mk 9 - Air Publication 4491J

Periodicals
Aeroplane Monthly February & March 1996, May & June 1998 and January 2004
Aircraft Illustrated August 1984
Air Pictorial January 1967
Fana de l'aviation No.202 & 205 (1986)
Flight, 9th March 1956
Flypast, January 2004
Modelaid International, May 1987
Scale Aircraft Modelling, 1979
Scale Aviation Modeller International, Vol.1 Iss.9, Vol.2 Iss.6, Vol.6 Iss.2 and Vol.9 Iss.10

INDEX

A&AEE Boscombe Down	17, 21 & 22, 27, 77
ADEN 30mm cannon	8, 11, 18, 28 & 29, 50
Air Defence Identification Zone	45
Air Fighting Development Squadron	29
Air Ministry	9,10, 11, 19
All-Weather Development Squadron	29
Armament Development Establishment, Enfield	50
Armstrong-Siddeley Sapphire	14, 18, 25, 28-30, 45, 49
Ashley, Isle of Wight	21
Avro 707A	11
Avro 707B	11
Beaton, George H Sqn Ldr	76
Birchfield, F E W Wg Cdr DFC	22
Blackburn Aircraft Co., Ltd	18
Boulton Paul P.111	11
Bristol Channel	21
Butterworth, Malaya	45
Carter, George	10
Central Fighter Establishment	29
Churchill, Winston	23
Confrontation Crisis	44-45
Controller Aircraft Release	21, 22
Crabbe, P G	10
Crew, E O Wg Cdr	21
Cuss J F	10
De Havilland 'Blue Jay' & Firestreak	21, 28-29, 50-51, 76
De Havilland DH.108	11, 12
De Havilland DH.110	9, 11, 12, 14, 16, 23
De Havilland DH.112 Sea Venom	12
De Havilland Gyron Junior	77
De Havilland Sea Vixen	30
De Havilland Venom	23, 25
Dick, David Sqn Ldr	21
English Electric Canberra	42, 44
English Electric Lightning	30-31, 43
Everest, P Col.	21
Exercise Dragon	27
Exercise Dyke	30
Exercise Malta ADEX 60	29
Exercise Pounce	30
Exercise Shiksha	43, 45
Exercise Sunbeam	29
Exercise Yeoman	28
Far East Air Force	45
Federation of Malaysia	45
Flax Bourton, Nr Bristol	19
Folland (IAI) Gnat	44
George Medal	17
Gloster Aircraft Company Ltd	8-10, 12, 14, 18-19, 22-23
Gloster Gladiator	27
Gloster Meteor	8, 22-23, 27, 29, 42, 45
Guided Weapons Development Squadron	51
Hawker Hunter	9, 23, 44-45
Husk, D I	10
James, I	10
Javelin Mobile Conversion Unit	23, 42
Jefferies Flt Lt	21
Johnson, R Col.	21
Kai Tak Airport, Hong Kong	45
Kunching, Borneo	44-45
Labuan, Borneo	44-45
Lawrence, Peter	18, 19
Lippisch, Alexander Dr	8
Lippisch P.13a	10
Luqa, Malta GC	29, 45
Lusaka. Zambia	43
Martin, R F 'Dicky' Wg Cdr	20 & 21
Mauser MK 213C	8, 50
Metrovik F.9	11
Ministry of Supply	9, 12, 14, 16, 17, 21-22, 25
Miller, M 'Dusty' Sqn Ldr	23
Moreton Valence	12, 14, 18, 22
Mutual Defence Aid Program	21
Nairobi, Kenya	43
Nassau Agreement	43
Ndola, Zambia	43
Nicosia, Cyprus	27, 29, 31, 43
North American F-86 Sabre	10 & 11, 29
North American F-100 Super Sabre	44
No.1 Guided Weapons Test Squadron, Valley	51
No.23 MU, Aldergrove	22
No.27 MU, Shawbury	45
No.228 Operational Conversion Unit	23, 27, 29
Operation Beware	21
Operation Readiness Platform	27
Paris Air Show	21
Penang, Malaya	45
Project	
P.228	8, 66
P.234	8, 66
P.238	66
P.240	66
P.248	8-9, 66
P.250	8-9, 66
P.258	66
P.259	9-10, 66
P.272	10, 66
P.275	10, 66
P.276	66
P.278 & 279	66
P.280	12, 66
P.315	66
P.316 to 319	66
P.322 to 326	67
P.347 & 348	67
P.350	67
P.356	67
P.359	67
P.364	67
P.368	67
P.370 to 372	67
P.376	67
P.382	67
P.505	67
Queen's Coronation Review	19
RAE Farnborough	11, 78
RAF Night Fighter Conversion Unit	29
RAF Regiment	43
RAF Stations	
Acklington	27
Akrotiri	29, 31, 43-44
Aldergrove	22
Binbrook	44
Brize Norton	17
Brüggen	23, 42
Cardington	20
Church Fenton	27
Coltishall	21, 27, 29-30
Duxford	29-30
Geilenkirchen	20, 31, 42
Gütersloh	43
Horsham St. Faith	23, 27, 29
Leconfield	27
Leeming	23, 27, 29
Leuchars	27, 28, 31
Middle St. George	29, 31, 42
Odiham	19, 25
Stradishall	25
Tengah	44-45
Turnhouse	27
Valley	50
Waterbeach	25, 29-30
Wattersham	29
West Malling	25, 29
West Raynham	29
Red Hawk, air-to-air missile	9
Rolls-Royce Avon	10 & 11
Ross, R J Flt Lt	21
Royal Norwegian Air Force	29
SBAC, Farnborough	18, 21, 27
Scott, P Sqn Ldr	21
Smith, Brian	18
Smith, Ian	43
Specification	
F.3/48	9-10
F.4/48	9-10, 12
F.43/46	8, 66
F.44/46	8, 66
N.14/49	12
N.40/46	11 & 12
T.118D	25
Squadron (RAF)	
No.3	42
No.5	31, 42-43
No.11	30, 42-43
No.19	43
No.20	45
No.23	27, 29-31
No.25	29-30
No.29	27, 31, 43-44
No.33	29, 31, 42
No.41	42
No.43	23
No.41	27, 29
No.46	25
No.60	44-45
No.64	29-30, 43-44
No.72	27
No.85	25, 29
No.87	23, 42
No.89	25
No.92	433
No.96	42
No.141	27
No.151	27
Steepholme	21
Street, P D C Sqn Ldr DSC	23
Supermarine Scimitar	30
Unilateral Declaration of Independence	43
Vickers Valetta	23
Vickers Valiant	20
Walker, Richard W	10
Walton, J Sqn Ldr	21
Waterton, W A Sqn Ldr AFC	13 & 14, 17 & 18, 20-21
White, H E Wg Cdr DFC, AFC	22, 27
Williamson Flt Lt	21
Worrall, Geoff	18, 21, 27
Zurakowski, Jan	18, 27
2nd Tactical Air Force	31, 42